Environmental Footprints and Eco-design of Products and Processes

Series Editor

Subramanian Senthilkannan Muthu, Chief Sustainability Officer, Green Story Inc., Tung Chung, Hong Kong

Indexed by Scopus

This series aims to broadly cover all the aspects related to environmental assessment of products, development of environmental and ecological indicators and eco-design of various products and processes. Below are the areas fall under the aims and scope of this series, but not limited to: Environmental Life Cycle Assessment; Social Life Cycle Assessment; Organizational and Product Carbon Footprints; Ecological, Energy and Water Footprints; Life cycle costing; Environmental and sustainable indicators; Environmental impact assessment methods and tools; Eco-design (sustainable design) aspects and tools; Biodegradation studies; Recycling; Solid waste management; Environmental and social audits; Green Purchasing and tools; Product environmental footprints; Environmental management standards and regulations; Eco-labels; Green Claims and green washing; Assessment of sustainability aspects.

Subramanian Senthilkannan Muthu
Editor

Energy Efficiency and Carbon Footprint Assessment

Case Studies

Springer

Editor
Subramanian Senthilkannan Muthu
Green Story Inc.
Tung Chung, Hong Kong

ISSN 2345-7651 ISSN 2345-766X (electronic)
Environmental Footprints and Eco-design of Products and Processes
ISBN 978-3-032-10908-8 ISBN 978-3-032-10909-5 (eBook)
https://doi.org/10.1007/978-3-032-10909-5

This Springer imprint is published by the registered company Springer Nature Switzerland AG
The registered company address is: Gewerbestrasse 11, 6330 Cham, Switzerland

This book is dedicated to:

The lotus feet of my beloved Lord Pazhaniandavar

My beloved late Father

My beloved Mother

My beloved Wife Karpagam and Daughters—Anu and Karthika

My beloved Brother

Last but not least
To everyone working in various industrial spaces to make the world, "SUSTAINABLE"

Contents

Contents

Introduction to the Book—Energy Efficiency and Carbon Footprint Assessment

Subramanian Senthilkannan Muthu

1 Introduction

This book examines one of the most pressing challenges of the modern era—the urgent need to reduce greenhouse gas emissions and improve sustainability across energy-intensive sectors. Industries such as textiles, construction, petrochemicals, transport, and energy production contribute significantly to global carbon emissions, both through direct operations and indirect supply chain activities. Improving energy efficiency and assessing carbon footprints are therefore essential strategies for mitigating climate change while maintaining industrial growth and competitiveness. This book provides a comprehensive analysis of the methods, technologies, and practices that support a transition towards low-carbon pathways. Key themes include renewable energy adoption, circular economy approaches, sustainable waste management, alternative fuels, and systematic assessment frameworks for evaluating and reporting emissions. The discussion extends beyond technological solutions to include governance structures, regulatory mechanisms, and best practices that enable large-scale implementation of decarbonization strategies.

By presenting cross-sectoral perspectives, practical case studies, and critical evaluations, the book aims to offer valuable insights for academics, policymakers, industry professionals, and sustainability practitioners. It emphasizes the role of energy efficiency as a cornerstone of industrial sustainability and highlights how carbon footprint assessment serves as a vital tool for informed decision-making, accountability, and alignment with global climate commitments.

The second chapter is about the Cotton waste management—Reducing industrial carbon footprint through sustainable practices. Cotton waste is one of the major problems with the textile industry, which contributes substantially to carbon

S. S. Muthu (✉)
Green Story Incl., Toronto, Canada
e-mail: drsskannanmuthu@gmail.com

S. S. Muthu (ed.), *Energy Efficiency and Carbon Footprint Assessment*, Environmental Footprints and Eco-design of Products and Processes,
https://doi.org/10.1007/978-3-032-10909-5_1

emissions and environmental degradation. Cotton waste is generated during various phases of textile production, like fiber processing, spinning, weaving, dyeing, and apparel manufacturing. If not managed properly, this waste causes landfill accumulation, water pollution, as well as greenhouse gas emissions. However, implementing sustainable waste management strategies can mitigate the textile sector's carbon footprint. This chapter explores the environmental impact of cotton waste and highlights innovative approaches to its management. It examines waste minimization techniques such as efficient processing methods, recycling, and upcycling initiatives. Additionally, it discusses the role of circular economy principles in reducing cotton waste and promoting resource efficiency. The potential for converting cotton waste into bio-based products, such as compost, biofuels, and regenerated fibers, is also analyzed. This chapter also evaluates policy frameworks and best industrial practices that encourage sustainable cotton waste management. Case studies of successful industrial applications are presented to demonstrate how companies can align with global sustainability goals while reducing their overall carbon footprint. By integrating eco-friendly solutions and waste valorization strategies, the textile industry can significantly decrease its environmental impact. This chapter aims to provide a comprehensive understanding of how cotton waste management contributes to a lower industrial carbon footprint and leads to a path for a more sustainable future.

The third chapter explains the Selection, Measurement, Evaluation and Reporting method of mitigation actions applied to sustainable urban freight transport sector. In alignment with global efforts to achieve the climate targets set by the Paris Agreement and mitigate environmental impacts, the urban freight transport (UFT) sector must undergo significant decarbonization to help reduce carbon footprint and consequently curb global temperature rise. Reducing greenhouse gas (GHG) emissions in UFT demands a multifaceted approach, including optimizing logistics and operational efficiency, transitioning to low-carbon and zero-emission transport modes, and adopting advanced fuel-efficient technologies. This chapter presents the Selection, Measurement, Evaluation and Reporting (SMER) method, a important tool for evaluating and tracking the effectiveness of these strategies. The SMER framework enables companies and policymakers to systematically measure and select sustainability initiatives, evaluate progress, transparently report results, and ensure the verification of emission reductions by enhancing the MRV method. Furthermore, it plays a critical role in supporting the implementation and assessment of Nationally Determined Contributions (NDCs) by providing a structured approach to evaluating three key dimensions: GHG emissions quantification, effectiveness of mitigation actions, and tracking of financial, technological, and capacity-building support. By establishing a robust SMER method, the UFT sector can enhance accountability, drive continuous improvement, and accelerate the transition toward a low-carbon economy.

The fourth chapter explains how the implementation of best sustainable practices in the urban freight transport sector can help the logistics sector achieve sustainable development goals. Urban logistics decision-makers face significant challenges in balancing operational efficiency with sustainability in urban freight transport (UFT). The increasing demand for fast deliveries, traffic congestion, greenhouse gas (GHG) emissions, and the need to minimize environmental impacts require the adoption of innovative and integrated strategies. In this context, it is essential to identify and implement best practices that optimize activity levels, promote the transition to low-carbon transport systems, reduce energy consumption, and enhance fuel efficiency by lowering carbon intensity. This chapter presents a comprehensive literature review aimed at mapping and analyzing the potential of best practices in UFT to contribute to the Sustainable Development Goals (SDGs) of the 2030 Agenda for Sustainable Development of the United Nations Organization. This chapter explores various approaches, including public policies for sustainable urban logistics, technological innovations such as fleet electrification and logistics process digitalization, and collaborative solutions like freight consolidation and the use of alternative transport modes. By gathering and evaluating these practices, this chapter seeks to provide valuable insights for policymakers, businesses, and researchers interested in driving the sustainable transformation of the urban freight transport sector.

The fifth chapter the potential of using ethanol as a vector for decarbonisation: An analysis based on SWOT matrix. The growing pressure for low-carbon solutions in maritime transport is accelerating the search for alternative fuels, especially in Brazilian coastal shipping. This chapter evaluates ethanol as a strategic decarbonization vector for the sector, considering the international regulatory context, the maturity of the national production chain, and the technical challenges involved. This chapter adopts a qualitative approach, based on literature review and document analysis, complemented by a SWOT matrix. Results indicate that ethanol offers significant advantages in terms of greenhouse gas emissions, existing infrastructure, and alignment with public policies. However, challenges such as fleet adaptation, the absence of specific regulations, and competition with other energy alternatives limit its adoption. This chapter contributes to a technical reflection on sustainable pathways in commercial shipping and proposes pathways to enable the use of biofuels in Brazilian coastal shipping.

The sixth chapter is about the critical analysis of biodiesel production: comparison of routes and potential of Distillers Corn Oil (DCO). This chapter presents a critical analysis of the feasibility of biodiesel production by comparing different production routes. Special emphasis is placed on the production of biodiesel from residual oil derived from corn ethanol production, known as Distillers Corn Oil (DCO). The analysis begins by defining what constitutes biodiesel and highlighting its advantages over fossil-derived diesel. It then addresses the need for sustainable processes to replace fossil energy sources and explores how circular economy principles and the United Nations Sustainable Development Goals (SDGs) can be advanced through the use of biodiesel produced from DCO via an enzymatic route. This chapter proceeds with an evaluation of existing reaction pathways—basic homogeneous, acidic homogeneous, CaO/SiO heterogeneous, KOH heterogeneous, and enzymatic—and discusses the advantages and disadvantages of the enzymatic route in particular. Finally, a comparative life cycle assessment of fossil diesel and biodiesel is presented.

The seventh chapter is about the role of Ethanol as an alternative for the sustainable expansion of electricity supply in Brazil. This chapter investigates the role of ethanol as an alternative for the sustainable expansion of electricity supply in Brazil.

The research contextualizes the growing demand for renewable energy sources and the challenges of greenhouse gas emissions. The objectives include analysing the viability of ethanol as a replacement for fossil fuels and assessing its environmental and economic impact. The methodology involves a literature review and analysis of production and consumption data. The main results indicate that ethanol is an alternative and still contributes to the reduction of emissions from the energy matrix, and its large-scale implementation will not face technical and economic barriers, with new generation technologies that have already overcome the challenges. This chapter concludes with recommendations for public policies that encourage the use of ethanol as a fuel in the electricity sector.

I take this opportunity to thank all the contributors for their earnest efforts to bring out this book successfully. I am sure readers of this book will find it very useful.

With best wishes,

Dr. Subramanian Senthilkannan Muthu

Cotton Waste Management—Reducing Industrial Carbon Footprint Through Sustainable Practices

Samar Gogari, Radha, and Ashok Pundir

Abstract Cotton waste is one of the major problems with the textile industry, which contributes substantially to carbon emissions and environmental degradation. Cotton waste is generated during various phases of textile production, like fiber processing, spinning, weaving, dyeing, and apparel manufacturing. If not managed properly, this waste causes landfill accumulation, water pollution, as well as greenhouse gas emissions. However, implementing sustainable waste management strategies can mitigate the textile sector's carbon footprint. This study explores the environmental impact of cotton waste and highlights innovative approaches to its management. It examines waste minimization techniques such as efficient processing methods, recycling, and upcycling initiatives. Additionally, it discusses the role of circular economy principles in reducing cotton waste and promoting resource efficiency. The potential for converting cotton waste into bio-based products, such as compost, biofuels, and regenerated fibers, is also analyzed. This study also evaluates policy frameworks and best industrial practices that encourage sustainable cotton waste management. Case studies of successful industrial applications are presented to demonstrate how companies can align with global sustainability goals while reducing their overall carbon footprint. By integrating eco-friendly solutions and waste valorization strategies, the textile industry can significantly decrease its environmental impact. This study aims to provide a comprehensive understanding of how cotton waste management contributes to a lower industrial carbon footprint and leads to a path for a more sustainable future.

Keywords Carbon footprint · Environmental impact · Waste management · Sustainability · Circular economy

---

S. Gogari · Radha (✉)
School of Biological and Environmental Sciences, Shoolini University of Biotechnology and Management Sciences, Solan, India
e-mail: radhuchauhan7002@gmail.com

A. Pundir
Department of Civil Engineering, Graphic Era (Deemed to be University), Dehradun, India

S. S. Muthu (ed.), *Energy Efficiency and Carbon Footprint Assessment*, Environmental Footprints and Eco-design of Products and Processes,
https://doi.org/10.1007/978-3-032-10909-5_2

1 Introduction

1.1 Global Consumption Trends and Environmental Impacts of Global Textile Industry

Within the last decade, the criticism faced by fashion and textile industry is a result of its negligence towards social and environmental sustainability. The industry has brought several environmental threats and intangible expenses related to pollution, water wastage, waste management and many others [1]. The dreadful consequences of fashion industry are not only widespread but also highly impactful on the environment, it is estimated that on an average this sector is responsible for contributing approximately 1.7 billion tons of carbon annually which is around 10% of global CO_2 emissions [1–3]. Additionally, fashion industry is the 2nd major consumer of water globally, using an estimated amount of 1.5 trillion litres of water annually for its various processes of textile dyeing and treatment. It is also a major contributor for industrial waste water, accounting approximately 20% industrial water pollution [4]. Synthetic textiles from these industries are a major contributor of microplastics, a major environmental concern accounting for about 35% of oceanic primary microplastics estimating around 0.19 million tons annually [1]. Additionally, textile waste has surged to drastic levels, surpassing 92 megatons annually. A large proportion of this waste including substantially high unsold garments are either dumped in landslides or incinerated [5, 6]. Rising demand of textiles and apparels is directly linked to environment degradation. Textile production in the world has escalated from 5.9 kg in the year 1975 to 13 kg in 2018, per year for a single person [7]. Fashion brands, currently have doubled their production in comparison to their production in 2000. This is mainly in response to the rising demand and consequently increasing global textile production approximately at a rate of 2% per year [8, 9]. This rise in production is the result of the surge in "fast fashion" model which is based on continual and rapid supply of cheap and trend-focused apparel. This model promotes spontaneous buying and habitual consumption, because of limited time and affordability [10, 11]. This model is beneficial over traditional retail as it takes advantage of online shopping platforms to reach maximum consumers, enabling their higher turnover of products [10]. The affordability of buying clothes has increased hence reducing the frequency of wearing them [6, 10, 12]. It is estimated that consumers now buy a new apparel every 5.5 days in the US [13]. There was a 40% increase in cloth purchase b from 1996 to 2012 across Europe [5, 14]. Studies have demonstrated that people in UK and Norway, impulsively buy clothes and discard them after minimal use [15–17]. Most of the data comes from the Global North, but similar trends for textile consumption are also emerging in the South because of the rapid urbanization, economic expansion and globalization of westernized fashion [18]. Sustainability frameworks like UN Sustainable Development Goals, particularly SDG-12 (Responsible Consumption and Production) and SDG-13 (Climate Action) have been put forth because of the increasing climate crisis and the need for adopting sustainability. This pressurizes all the stakeholders within the textile and fashion industries to accept and practice more environmentally responsible and sustainable approaches [1]. This book chapter examines how cotton waste from the textile industry significantly contributes to environmental challenges like carbon emissions, water usage, pollution, and landfill accumulation. It analyzes the sources and types of cotton and textile waste, explores the life cycle environmental impact (including greenhouse gas emissions, water and chemical use), and assesses both mechanical and chemical recycling methods alongside energy recovery options such as bioethanol and biogas. The chapter emphasizes innovative strategies like artificial intelligence (AI) for optimizing waste management and the integration of circular economy principles, illustrating that these efforts, when combined with effective policy frameworks, can reduce the industrial carbon footprint, promote sustainability, and enable the creation of valuable products from cotton waste.

1.2 Cotton as a Contributor in Global Carbon Emissions

Due to its widespread and extensive applications across diverse industries like fashion, textiles and healthcare, continues to be the most prevalent and dominant natural fiber mainly because of its distinctive physical and chemical properties [19]. Apart from its primary uses, cotton holds an abundant economic value for by-products like cottonseed oil and livestock feed, thereby contributing to both national and international economies worldwide. This made cotton earn its status as "white gold" [20]. In countries like India, cotton plays a crucial role in driving the economy, where it provides livelihoods to approximately 60 million people by facilitating employment across the entire cotton value chain like cultivation, processing and textile production [21]. Despite its economic significance, cotton production and processing raise exorbitantly high environmental concerns, particularly in the fast fashion industries. The fast fashion concept has alleviated the consumption of clothing and has raised concerns about environmental issues and climate change [17, 22]. Consequently, the discussions about the textile industry and its intensifying carbon footprint is exacerbating, hence demanding to alleviate emissions by taking some mitigation measures [23–26]. It is estimated that with rising demand for cotton, the global lint production will rise by 1.3% per year, further growing to 29 million metric tons by 2033 [27]. Without widespread acceptance and adoption of sustainable agricultural practices, this anticipated growth will exacerbate cotton's carbon footprint on the environment. To estimate environmental impacts at every stage of cotton value chain, Life Cycle Assessment (LCA) is the most commonly implemented technique in majority of carbon footprint analyses [28]. Spatial and temporal trends of emission were given by a regional LCA carried out in China between 2004 and 2018 [29]. The carbon footprint of cotton lint was found to range between 2.49 and 3.73 kg carbon dioxide equivalent (CO_2e) per kilogram, influenced highly by input-intensive practices like irrigation, use of pesticides and fertilizers and application of mulches. Various reports from corporates and organizations have quantified and presented data for the carbon footprint of cotton across various countries like India, China, US, Australia and selected regions of Africa [30, 31]. The studies are not just limited to investigation of carbon emissions at farm-level, instead there are several studies which evaluate carbon emissions across the entire cotton value chain. In one of the studies it was found that the lifelong emissions of end-use cotton products like t-shirts and denims were much more less than the agricultural phase of production which accounts for significant carbon emissions [32, 33]. A farm-level study was conducted to evaluate the carbon emissions from cotton farms and to assess effectiveness of mitigation measures for reducing greenhouse gas (GHG) emissions. The results in this study indicated that nitrogen-based fertilizers are the major contributors to these emissions. Therefore it is clear that clean energy-based farming and targeted-fertilizer application can reduce GHGs by around 20% [34]. For such studies, cotton ginning is often excluded due to limited data availability and the functional unit being one ton of cotton yield from the farm. The two predominant GHGs emitted in these systems are carbon dioxide (CO_2) and nitrous oxide (N_2O), which are a result of fuel-burning and application of N-based fertilizers respectively. Agricultural inputs like fertilizers, pesticide, fuel and electricity show regional variations because of differences in agricultural practices and machineries [35, 36].

1.3 Objectives and Scope of the Study

This study provides a comprehensive and detailed analysis of the various environmental impacts associated with cotton farming and textile waste generated after cotton processing. The main focus lies on the carbon footprint analysis across the entire value chain of cotton. The primary objective is the identification of GHG emission sources both from agricultural and industrial processes and thus exploring various mitigation measures or strategies relevant to circular economy framework. The study also evaluates the classification, generation and LCA of cotton waste, from agricultural residues and industrial by-products to pre and post-consumer textile waste. Further, the study provides insights on recycling technologies, renewable energy recovery and AI-powered innovations for sustainable waste management. By combining environmental assessments with emerging policies, advanced technologies and market trends, this study aims to guide policy-makers, researchers and industry-stakeholders in developing sustainable strategies to reduce emissions, save resources and promote the upcycling of cotton waste into high-value efficient products.

2 Cotton Textile Waste—Sources and Types

2.1 Types of Cotton Waste

Cotton farming and its processing both generate substantial amount of waste that can be broadly classified into two types based on the source: Agricultural residues and industrial by-products.

2.1.1 Post-harvest Cotton Waste (Stalks and Roots)

Post-harvest, cotton farms are left with significant amount of biomass which includes stalks, roots and leaves. Naturally, cotton plants undergo a process of defoliation before being harvested, hence majority of the leaves fall-off naturally, only roots and stalks which remain are left in the soil [37]. Traditional practices of ploughing or re-incorporating these residues back into the soil, are energy consuming and sometimes may degrade the structure of soil [37, 38]. Therefore, when these agricultural residues just remain on the farms, they are agronomically considered as low value biomass. But these waste residues exhibit huge potential for generating bioenergy, as the calorific content in cotton stalks ranging between 14.5 [39] and 17.1–18.1 MJ/kg [38], is similar to that of wood which ranges from 17.4 to 18.6 MJ/kg [40]. These characteristic features of cotton stalks help cotton-producing countries like India, China, Brazil, Australia, Pakistan and Turkey in producing significant amount of bioenergy from these residues [41]. Though 80% of the global cotton production in 2015–16 came from top 5 producers, namely China, India, US, Pakistan and Brazil, more than half of this production is contributed by China and India alone [39]. Due to their high agricultural production, such countries the contribution of plant-waste (stalks, leaves and hulls) is very high in such countries [42].

2.1.2 Industrial Cotton Waste (Gin Trash)

Every stage of cotton textile garment production generates cotton waste. For reusing and recycling this waste it is important to have a knowledge about different types, quality as well as amount of cotton waste generated within the supply chain [39]. An excessive amount of biomass remains decaying on the farms after cotton is harvested. Thereafter, during the process of ginning around 35% of cottonseed is separated from cotton fibers and other residues. Although cottonseed accounts for about 85% of non-lint residue, around 15% remains unutilized even after using it for oil production and animal feeds. However this 15% fraction has many potential applications like ethanol production, for oil-spill remediation or fertilizers [39]. Ginning is a process of separating cotton fibers from seeds. Approximately, 10% of the processed material during the ginning process is discarded as cotton gin trash (CGT) which is a heterogeneous mixture of burrs, stems, dirt, lint, leaf fragments and seeds [39]. Post-ginning, the fiber is baled, then undergoes bale opening, blending, cleaning and carding, before being sent for spinning. Here the carding process solely is responsible for 4.7% of waste generated. Interestingly, the recoverable fiber content in the flat and undercard process is 65% and 56% respectively, indicating recovery potential [43]. The next step of drawing includes blending and alignment of fibers, followed by combing process to remove short fibers and to produce consistent, fine yarn. During this combing step, approximately 20% of the input is waste and sent for disposal [44]. Depending on the machinery, approximately, 5% yarn waste is generated during spinning [45]. Depending upon the complexity of pattern and the type of loom during weaving process, the generation of waste percentage varies. For instance 5–7.5% of cotton is rejected and discarded due to selvage, knotting and other defects in production [45]. Fabric cutting and sewing in the garment manufacture approximately results in 15% of fabric waste, even after precision cutting methods [45, 46]. Moreover, during the manufacturing of T-shirts, approximately 6.5% of waste is generated only during sewing errors and quality control assessments [46]. Table 1 gives a comparative overview of pre- and post-consumer textile waste in the textile supply chain.

2.2 *Types of Textile Waste*

Global textile waste has increased to alarming levels because of rising production and consumption by fast fashion industry. In the past, Western nations used to export waste garments to developing countries particularly Africa, for managing this waste [47]. However, many developing countries have started imposing ban on importing textile waste with a purpose of protecting and preserving local textile sectors as well as with the fear of threat to domestic production [10, 47]. For better understanding of its origin and management, textile waste is classified into pre-consumer and post-consumer textile wastes.

2.2.1 Pre-consumer Textile Waste

Generated during textile and garment production and comprising fiber, yarn, damaged yardage, waste fabric as well as sample garments, pre-consumer textile waste is often referred to as production waste. Studies estimated that despite being the most resource-intensive, approximately 15% of the fabric during garment production is discarded [17]. Some studies indicate that textile waste during manufacturing has the potential to reach 25–30% [48]. Factors like garment design, fabric width, pattern layout efficiency and surface design (one way prints), all influence the percentage of textile waste. Waste increases during the cutting stage because of inefficient pattern design and defects in garment assembly [48]. Production waste will continue to grow because of the blooming fashion industry, hence to mitigate this issue, reducing the rate of production as well as enhancement in manufacturing design are sustainable

Table 1 Difference between post-harvest and industrial cotton waste

Aspects	Post-harvest cotton waste	Industrial cotton waste	References
Definition	Biomass remaining on fields after cotton harvesting, including stalks, roots, and leaves	Waste generated at various stages of cotton processing, including ginning, spinning, weaving, and garment manufacturing	[37, 39]
Primary components	Stalks, roots, defoliated leaves	Cotton gin trash (burrs, stems, dirt, lint), cottonseed waste, fiber waste, yarn waste, fabric scraps, sewing rejects	[39, 43, 44, 46]
Source location	Agricultural fields	Cotton ginning units, spinning mills, weaving units, garment factories	[38, 45]
Agronomic value	Traditionally considered low-value biomass, though often ploughed back into soil	Limited agronomic value; mostly disposed or recycled depending on quality and form	[37, 39]
Energy potential	High calorific value (14.5–18.1 MJ/kg), comparable to wood; suitable for bioenergy	Some potential for bioenergy (e.g., cottonseed residues for ethanol, compost)	[38–40]
Volume of waste	Large quantities in top producers like China, India, Brazil, and Pakistan due to high production volumes	Around 35% cottonseed separated during ginning; 10% gin trash; ~ 20% loss in combing; 15% cutting waste; 5–7.5% weaving waste	[39, 43–46]
Recycling/reuse options	Bioenergy production, composting, mulching	Fiber recovery, animal feed, oil extraction, ethanol, padding, stuffing, oil spill remediation	[39, 41, 45]
Environmental concerns	Energy-intensive re-incorporation may degrade soil structure	Waste contributes to landfill burden; QC waste is often discarded	[11, 38]
Recovery potential	High if utilized for bioenergy or composting	~ 65% recoverable fiber in carding waste; potential in combing, spinning, and weaving rejections	[43, 44]
Key producers	China, India, Brazil, Australia, Pakistan, Turkey	Countries with significant cotton textile industries (India, China, Bangladesh, Vietnam)	[39, 42]

solutions [49]. An emerging area of concern within pre-consumer waste is deadstock, which consists of new but unsold or returned garments, often discarded as waste. In 2016, it was reported that in the EU, only one third of imported clothing was sold at full price, with the rest sold at discounts or not at all [50]. Same trend was observed in Netherlands, wherein around 21 million garments (6.5% of total production) remained unsold in 2015 [17]. Major brands worldwide have been facing criticism for the negligent and mismanaged handling of their deadstocks. One primary example of H&M, wherein this reputed brand was reported to hold around $4.3 billion in unsold garments [51], out of which some stock was burned in Danish waste-to-energy plants [52, 53]. Similarly, Burberry incinerated £28.6 million worth of garment products in 2017 and £90 million over the period of 5 years [17, 54]. Though incineration helps in recovering some percentage of energy at the same time it is responsible for releasing pollutants and GHG emissions. Also it is less environmentally favorable compared to reuse or recycling [6]. Significant energy input, water, and chemicals required to manufacture these unsold garments are the biggest environmental concerns other than textile waste [49]. Monitoring pre-consumer textile waste is difficult because private disposal agreements hide data on waste volume, material content and end destination of the waste [55]. While some textile producers or brands practice recycling or donate the residual stock, others discard and destroy unsold garments to protect brand reputation [56].

2.2.2 Post-consumer Textile Waste

Post-consumer textile waste includes clothes and apparels discarded after consumer use. Out of the estimated amount of 150 billion garments in 2012, around 60% were discarded within a few year span [9, 57]. The average duration of use for commonly purchased garments like T-shirts, knit shirts, and pants varied from only 3.1 to 3.5 years across countries like China, Italy, Japan, Germany, UK, and USA [58]. There has been a 40% rise in textile waste sent to landfills of USA in just a single decade from 1999 to 2009, owing to this trend of short using duration of garments and rising purchase [45]. 22% of municipal waste is constituted with textiles [59]. 73% of produced textiles (approximately 39 gigatons) in 2015, were landfilled at end-of-life [6]. On an average 30 kg of textile waste both in UK and USA [60], 27 kg in Australia [17], 16 kg in Denmark [17], and 13 kg in Finland, per capita, is generated, annually [6]. Despite these extremely alarming figures, there is no rise in recycling rate, because in the same year (2015), only 15% of this waste was separated for recycle, out of which only 1% was recycled in a closed-loop system and most of these textile waste was downcycled into lower value products like stuffing and wiping cloth. About 1.1 million metric tons was lost while collecting and processing [6]. Escalating garment recycling along with reducing the production and consumption rate will help in achieving a circular textile economy [6, 49]. Table 2 gives a comparative overview of pre- and post-consumer textile waste in the textile supply chain.

Table 2 Comparative analysis of pre-consumer and post-consumer textile waste

Aspect	Pre-consumer textile waste	Post-consumer textile waste	References
Definition	Waste generated during textile and garment production before available for purchase	Waste generated after consumer use of garments	[9, 17]
Examples	Fiber, yarn, damaged fabric, cuttings, sample garments, deadstock	Used clothing, household textiles, discarded garments	[6, 48]
Causes	Inefficient cutting, poor pattern layout, design flaws, overproduction, defective assembly	Short usage lifespan, fast fashion, rising consumption	[45, 48, 58]
Waste proportion	15–30% of fabric wasted during garment production	~ 73% of produced textiles landfilled; < 1% closed-loop recycled	[6, 17, 48]
Influencing factors	Garment design, fabric width, surface design (e.g., one-way prints), pattern layout efficiency	Consumer behavior, low utilization rate, limited recycling systems	[6, 49]
Environmental impact	High resource input; incineration of deadstock releases pollutants and GHGs	Landfill burden, GHG emissions, downcycling of materials	[6, 49, 52]
Monitoring challenges	Difficult to track due to private disposal contracts obscuring data	Weak infrastructure for collection and sorting	[6, 55]
Industry practices	Brands may recycle, donate, or destroy excess stock to protect brand value	Consumers may donate, sell, repurpose, or store garments	[17, 51, 56]
Key statistics	H&M: $4.3B in unsold garments; Burberry incinerated £90M worth of products over 5 years	Annual per capita textile waste: UK/USA—30 kg; Australia—27 kg; Denmark—16 kg; Finland—13 kg	[6, 17, 51]
Sustainable solutions	Reduce overproduction, improve design efficiency, manage deadstock responsibly	Extend garment use phase, improve recycling, reduce consumption, support circular systems	[6, 49]

2.3 Global Statistics of Textile Waste

Lately, post-consumer waste is being monitored and documented worldwide but Europe has documented it primarily thus giving an outline of trend for textile waste worldwide. UK was responsible for generating around 2.35 million tons of textile waste in 2004, out of which 74% was landfilled, 13% was sent for incineration and only 13% was recycled [60]. Around one-fourth of this waste was collected via charitable donation centers [61]. Over 15–20% of the discarded textiles in Europe consisting of reused and downcycled clothes are usually exported to developing countries for their management [45]. Collection rates vary across different European nations. Germany recovers about 70% of its disposed textiles, whereas Denmark can only recover 40.5% recovery rest 38.5% going for incineration [62]. Regional studies in Nordic countries have estimated that around 80% of textile waste is recyclable, but lack of efficient technologies for recycling and high operational costs, limit the circular textile practices [45], a survey study of more than 6000 consumers across 6 countries including China, Japan, Germany, Italy, UK and US, revealed that 38% donate their used clothes, 26% discard them and only 15% repurpose their clothes, while the remaining either give away, sell or store their old garments [58]. The US lacks complete and systematic data on the fate of pre- and post-consumer textile cotton waste, even though 98% of the apparel is imported [45]. Out of the 2.7 million tons of cotton apparel imported in US, only 70% was sold, while the remaining were either funneled into secondary channels or destroyed [63]. It is estimated that out of the sold cotton apparels, 15% go for recycle, 19% for incineration for energy recovery and major part of 60% goes to landfills [45]. And that too from the recycled stock, 45% is sold in second-hand markets, 30% converted and tailored into cleaning rags, 20% for insulation or stuffing and rest 3% goes for donation in disaster relief [45]. 71% of the 790,000 metric tons of second-hand garments is exported overseas, which means only 32% of recycled textiles are destined for foreign resale or downcycling [45]. Approximately, one third of this waste is cotton which accounts to 260,000 metric tons of cotton which is shipped overseas [45].

3 Environmental Footprint of Cotton Waste

The modern textile industry is circling around a linear model, utilizing significant resources and minimal recycling and thus having profound environmental consequences. This wasteful approach also exerts pressure on global resources and contributes to environmental degradation by releasing harmful chemicals and shedding microfibers during washing [64, 65]. Synthetic materials like polyester, owing to their large-scale use as well as high demand, depend heavily on fossil fuels and non-renewable sources of energy. For example, plastic-based textile fibers production is estimated to consume 342 million barrels of oil per year [66, 67]. Cotton and its allied sectors have been playing a significant role in boosting the economies of various countries worldwide. But along with this development what accompany are the disastrous effect on the environment [68]. One of the primary concerns in cotton cultivation is the widespread use of chemical inputs such as fertilizers and pesticides, which contribute substantially to non-point source pollution [68]. Additionally, cotton farming demands puts a substantial stress on water resources, intensifying stress on water-scarce regions [69]. In the industrial sector, textile production is a major source of GHG emissions and is responsible for introducing chemical pollutants like dyes, softeners, and surfactants into aquatic ecosystems [17, 70]. These issues are amplified by the increasing demand for cotton, majorly fueled by fast fashion trends that stimulate overconsumption [17, 22]. Moreover. The environmental impacts of cotton production and consumption vary across diverse regions, affected by local agricultural systems, manufacturing procedures and consumer behaviors. These all factors are frequently accessed during LCA [33]. Further detailed environmental effects of cotton production are detailed further.

3.1 Environmental Effects

3.1.1 Pesticide Contamination

Conventional cotton cultivation requires substantial chemical inputs with approximately 200,000 tons of pesticides and 8 million tons of fertilizers per year [71]. Despite occupying only 2.5% of arable land, cotton cultivation alone accounts for 16% of global pesticide use [72]. In India, cotton farming consumes 50% of all pesticides, often causing severe health consequences for farmers [64, 73]. The average pesticide application rate was 1.57 kg/ha in 2015, with around 43.6 million kilograms of active ingredients [74]. The majority of this pesticide use is highly concentrated in five countries, with Brazil exhibiting rates of application nearly four times the global average [74]. While insecticide application in cotton decreased from 19% of global pesticide use in 2000 to 14.8% in 2010, whereas herbicide use remained relatively stable and unchanged during the same period [75]. The reduction in insecticide application is majorly due to introduction of genetically modified Bt cotton, integrated pest management strategies, and regulatory prohibition on conventional pesticides [76, 77]. In Argentina it was observed that there was 50% reduction in insecticide usage and better crop yields compared to conventional cotton because of the introduction of Bt cotton [78]. On the other hand, some regions such as India and Pakistan saw increased use of insecticide because of pest outbreaks like mealy bugs and pink bollworms [75]. Use of pesticides in cotton cultivation is particularly harmful for soil organisms like earthworms [79] as well as exposure to chemicals causes health risks to farmers like respiratory and skin infections [80, 81].

3.1.2 Fertilizer Contamination

Cotton cultivation on an average demands 150 kg/ha of nitrogen and phosphorous-based fertilizers which is somewhat similar to major food crops like rice, maize and sugarcane [82]. Fertilizer application rates tend to be lower in resource-limited conditions and higher in regions with well-developed irrigation and advanced machinery [83]. Environmental outcomes from fertilizer application are highly influenced by factors like yield potential and the income of farmers, along with intensity of mechanization and local conditions [84]. In heavy clay soils in cotton systems under irrigation, only 17–40% of applied nitrogen is utilized by plants, whereas 47–55% escapes into the atmosphere and 7–10% remains in the soil [85]. Ideally fertilizer application should increase nutrient uptake while reducing the losses to the environment. However, contrary to this nearly half of the cotton-producing countries overuse fertilizers and thus there are elevated nitrogen levels in groundwater and surface water due to nutrient runoff [86]. On the contrary, 40% of countries underuse fertilizers, resulting in nutrient deficiencies that deteriorate soil health and reduce long-term agricultural sustainability [82]. Textile production is also responsible for severe water pollution. A prime example is the Citarum River in Indonesia, where over 200 factories release dyes and other chemicals into the water, causing a threat to ecology [87].

3.1.3 Greenhouse Gas (GHG) Emissions

Other than the chemical pollution, the textile as well as cotton industry also contributes significantly to greenhouse gases and these gases are the major contributors to global warming. GHG emissions across the cotton textile value chain are characteristically represented by CO_2-equivalent (CO_2e) values. GHG emissions in cotton farming have both direct and indirect sources. Direct emissions are from on-farm activities like fuel burning and N-based fertilizer application. Indirect N_2O emissions come from ammonia volatilization and nitrate leaching, with emission quantification typically following IPCC guidelines [88]. Indirect emissions are largely associated with the upstream production of inputs. Fertilizer production, especially nitrogen fertilizer, exhibits marked regional emission disparities. China, relying heavily on coal, registers approximately 8.3 t CO_2e per ton of N fertilizer which is substantially higher than European figures [89–91]. Enhancing NUE aligns country-specific cotton NUE with national agricultural averages, with possible interventions including integrated soil-crop management [92], the 4R approach [93], and knowledge-driven N application [94, 95]. The carbon footprints reported vary between 3 and 62 kg CO_2e per product corresponding to 50–872 g per use, which closely aligns with the weight of the garment [96]. The stages with highest emissions differ from one study to another. Many assessments have demonstrated that manufacturing processes like spinning, weaving and wet processing are responsible for highest emissions [97–100], whereas other studies attribute higher impacts to consumer activities, particularly laundering [101, 102]. In the year 2015, textile production contributed around 1.2 billion CO_2e, lot more than the combined emissions done by maritime shipping and international fights [103]. Additionally, gradual and steady shift toward air transport from maritime shipping are adding more to the carbon intensity of textile distribution [104]. The GHG emissions of textiles are remarkably high, with each tons of fabric emitting 17 tons of CO_2e, in comparison with the 3.5 tons for plastics and less than 1 ton for paper [105]. Moreover, consumer practices like washing and drying clothes are responsible for an estimated 120 million tons of CO_2e annually [106, 107]. Determining whether production phase or use phase contribute more to overall GHG emissions is crucial for developing targeted sustainability policies [101].

3.1.4 Over-Consumption of Water

Water consumption is equally high across the entire textile value chain. This industry consumes approximately 93 billion m^3 of water annually, which is equivalent to 4% of global freshwater consumption, largely driven by cotton cultivation and fabric production [108]. Several major cotton-producing nations, including China, India and US are currently dealing with water scarcity [109]. Additionally, washing machines are responsible for a surplus 20 billion m^3 of water globally, every year [105]. Some LCAs estimate raw consumption of volumes of water whereas others implement scarcity-weighted indicators to quantify the water extraction in a particular region [110]. Cotton's heavy reliance on water makes it a major contributor to water stress in various agricultural regions. Scarcity-adjusted assessments demonstrate that 97–98% of total water consumption and impact originate from cotton farming, with household laundering contributing marginally [104, 110]. Approximately 50% of cotton fields are irrigated with blue water sources, instead of relying solely on rainfall [111]. Cotton can account for up to 726 mm of global blue water consumption, per year [112], making cotton one of the top 5 crops responsible for unsustainable blue water use. Because of this, significant impacts are seen in regions like Indus, Aral, Syr-Darya, Mississippi and Tigris-Euphrates river basins, where an estimated 300 million people face seasonal water shortages [112]. The global average water footprint for seed cotton production is significantly higher than most of the food crops including cereals and vegetables standing at 3588 m^3 per tons [113]. Seasonal water requirement for cotton range between 700 and 1200 mm, influenced by factors like cultivar, climate, and season length [29]. From 2000 to 2010, regions having less intensive agricultural systems showed an increase in water consumption per unit of cotton produced, whereas regions with medium and highly advanced agricultural systems witnessed enhanced water-use efficiency (WUE) due to technological advancements like drip irrigation [114]. Australia reported a 40% improvement in cotton WUE from 1988 to 2010, and the US improved its irrigation water efficiency from 0.05 to 0.20 kg/m^3 between 1980 and 2015 [114]. Contrastingly, Central Asia experienced loss in efficiency after 1990, largely because of soil degradation and less use of fertilizer [115]. In the coming years, although there will be an increase in cotton yields as well as WUE, water-stress is still going to prevail because of rising threats associated with climate change. This will potentially render cotton cultivation unsustainable in many countries of the world [116–118].

3.1.5 Eutrophication and Microplastic Pollution

Various stages of cotton textile LCA show eutrophication impacts, which is nothing but the heavy nutrient load in the aquatic systems [98]. There are various contributors to eutrophication varying across LCAs. Fertilizer application during cultivation can account for up to 83% of total emissions [98, 99]. Detergent discharge from washing machines and other laundering activities have increased and reached at 91% in certain studies [102]. Wastewater which is released from wet processing in cotton textiles also contributes substantially to eutrophication [119]. Usually, eutrophication impact profiles reflect patterns of GHG emission because of the same emission source, for example, nitrogen oxides from energy generation release GHGs along with their impacts on freshwater resulting in eutrophication [102]. Apart from this, microplastics from textiles are an increasing concern. Approximately 500,000 tons of plastic microfibers are released into the ocean annually in the process of washing synthetic textiles. This number is equivalent to around 50 billion plastic bottles [65, 120].

3.2 Life Cycle Assessment of Cotton Textile Industries

Although, direct impacts of cotton cultivation are extensively recognized, other impacts which arise due to various processes in entire cotton value chain. The LCA methodology provides a comprehensive framework to estimate and evaluate the environmental impacts right from the very first stage of raw material extraction, followed by manufacturing, use and till the last stage of disposal [121]. The life cycle of cotton textiles includes a sequence of consecutive processes, beginning with the production of agricultural inputs and cultivation, followed by yarn and fabric manufacturing, final product assembly, retail distribution, consumer usage, and ultimately, end-of-life disposal or recycling [35, 122–124]. Each of these stages usually occurs across different geographic locations, requiring continuous transportation [104]. After harvesting, cotton undergoes the process of ginning, where lint is compressed into bales for further processing [35]. Yarn production involves processes like blowing, carding, drawing, roving, and spinning [122]. Weaving or knitting facilitate fabric formation, whereas nonwoven varieties are produced from bonded fiber webs [124]. Wet processes like bleaching, dyeing, and washing are implemented for fabric finishing. Final garment assembly comprises cutting, sewing, pressing, and packaging before retail [124]. The phase of consumer use involves textile consumption activities of laundering, drying, and ironing [125]. Textile waste keeps emerging throughout the life cycle and is disposed by various means like landfilling or incineration as well as it is either reused or recycled. Textile waste recycling may follow a closed-loop approach by reintegrating fibers into the textile industry, or an open-loop approach by repurposing it into alternative products like biochar [45]. Majority of published LCAs evaluate all the processes from cotton cultivation to wet finishing, while very limited research extends to the stage of consumer use or end-of-life assessment [124]. Yu and Yang [126] conducted a study to estimate the life-cycle GHG emissions of global cotton production in year 2020 to evaluate the possible climate benefits of several field management strategies. By implementing LCA method, they demarcated the system borderline for cotton production, separating it into two main stages, initial stage being actual cotton cultivation processes or field operations on the

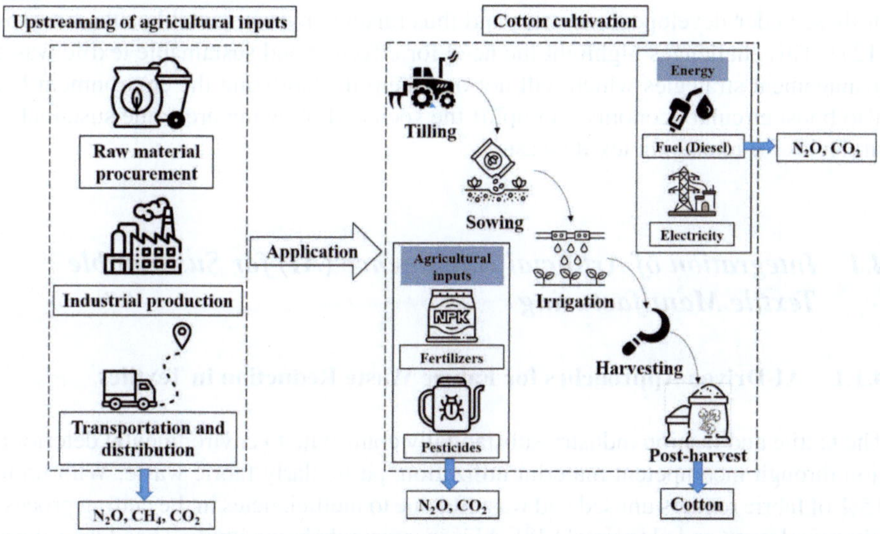

Fig. 1 Schematic representation of system boundary defined for the LCA method employed for cotton production. *Source* Yu and Yang [126]

farm, including ploughing, seeding, growing, harvesting, as well as other post-harvest activities. The 2nd phase comprises upstream processes, involving manufacture and transport of agricultural inputs like pesticides, fertilizer and fuels. CO_2 as well as N_2O were identified as the major GHGs released from these processes. Functional unit was demarcated as 1 ton of cotton produced on the agricultural field. Processes apart from the farming stage, like ginning, were excluded because of data limitations [126]. Figure 1 gives a schematic representation of system boundary defined for the LCA method employed for cotton production.

4 Sustainable Waste Management Strategies

Textile waste being a rising concern has various environmental and socio-economic impacts. Commonly this waste is either incinerated or landfilled both of which have negative impacts on the environment by releasing microplastics in the water air and soil. Economically, textile waste causes a huge loss to the industry. It is estimated that due to underutilization of clothes and lack of efficient recycling strategies, around $500 billion losses are incurred by the textile industry [127]. Socially textile waste exacerbates global inequalities, wherein used garments from developed nations are exported to under-developed or developing countries, where they are sorted into categories by low-wage labors and then sold in second-hand markets [127]. Again while sorting, when the garments which are not suitable for resale are discarded and disposed, further increasing the municipal solid waste (MSW) of these under-developed countries and thus raising environmental health concerns [127]. This challenges highlight the need for effective and sustainable textile waste management strategies which will not only help in stabilizing the environment but also boost circular economy and uplift the society. Following are some sustainable strategies for managing textile waste.

4.1 Integration of Artificial Intelligence (AI) for Sustainable Textile Manufacturing

4.1.1 AI-Driven Approaches for Fabric Waste Reduction in Textiles

The textile and fashion industry substantially contribute to environmental deterioration through incompetent material utilization, particularly fabric waste. With up to 15% of fabric renders unused and wasteful due to inefficiencies in the cutting process like poor layout and planning [128]. AI is increasingly being implied to address these inadequacies by optimizing production planning and material utilization. Machine learning has been found useful in pattern design particularly. Shelly Xu Design (SXD), developed an AI-based system which reconfigures traditional patterns into interlocking designs, thereby maximizing fabric usage and thereby minimizing fabric wastage. This innovation enabled 46% reduction in fabric waste and conserving as much as 41,000 m of fabric in one season [129]. By adapting to material limitations, product specifications, and surplus inventory, these tools surpass conventional nesting software. Tools like "IntelloCut" are aiding real-time optimization of fabric cutting and distribution in the manufacturing phase. IntelloCut helped Duc Hanh Garment JSC in reducing waste by 1.7% and also by decreasing material costs by 2.2%. It does precise estimation of fabric requirements and refine layout strategies accordingly [130]. AI is also restructuring demand forecasting practices. Nike's implication of the AI startup "Celect" enhanced the brand's forecasting accuracy by 85%, thereby help reduce overproduction and minimalize inventory-related fabric waste [131]. These advancements demonstrate the rising use of AI in supporting material efficiency and advancing circular economy practices within the industry.

4.1.2 Application of AI in Conserving Water During Textile Processing

Water consumption and pollution are the most concerning environmental challenges in the textile industry, which contribute approximately 20% of global industrial pollution of water. With significant water footprint in textile production, cotton fabric approximately requires 10,000 L of water while wool requires 170,000 L [132]. AI technologies are being implemented to alleviate these environmental impacts through enhanced process optimization and water management. AI-powered dyeing systems are one of the most effective innovative technique in this domain, where these systems minimize the need for re-dying, by adjusting dye formulations, based on previous data and current environmental parameters like pH level, temperature and fabric properties. Early defect detection made possible by Ekoten Tekstil's application of AI-enabled quality monitoring tools resulted in a significant water-use reduction and the frequency of re-processing [133]. Additionally, by minimizing chemical overuse, these systems help in lowering wastewater discharges. Furthermore, AI is developing smart water management systems. By integrating Internet of Things (IoT) sensors with AI-powered platforms, textiles can modulate water usage in real time, detect pipe-leaks and thereby evaluate possibilities to reuse water [132]. In some applications, AI helps in on-site water quality assessment, to check as to when recycling can be done with safety and feasibility, facilitating the implementation of closed-loop water systems. In agriculture, AI-powered irrigation systems are based on the utilization of real-time data on weather and soil moisture for effective optimization of water use in cotton cultivation enabling more sustainable management of resources [132].

4.1.3 AI-Powered Efficiency in Textile Production

Because spinning, dying, drying and finishing are power consuming processes, the production of textiles is a tremendously energy-intensive process. When energy use is not effectively regulated, these processes not only raise operating expenses but also drastically escalate GHG emissions [128]. Because AI technologies provide real-time insights and scalable optimization capabilities, they are being applied on a large scale to improve energy efficiency across production value chain. Modern AI-based Energy Management System (EMS), like Honeywell Forge and Schneider Electric's EcoStruxure, are now widely used for controlling energy loads and optimize production schedules. Manufacturers are able to detect inefficiencies like idle machinery, excessive HVAC use and outdated machinery with the use of these solutions [134]. Another advantage of AI integration is predictive maintenance, where AI-powered systems can identify early indicators of machine wear and imbalance by continuous evaluation of data streams from embedded sensors. This helps manufacturers prevent machine failure and make sure that the equipment works efficiently and smoothly. Smartex.ai uses image recognition and machine learning to track exactly how knitting machines operate, allowing for accurate maintenance scheduling and reducing energy loss from defective operations [132]. Additionally, AI demonstrates tremendous potential for optimizing supplementary systems, like compressed air management and steam generation which are frequently overlooked despite consuming substantially high energy. This can further lead to cumulative cost-efficiency [128]. By. 2030, AI-enabled energy efficiency technologies could reduce industrial emissions worldwide by 10%. Even relatively minimal energy savings have a significant impact on cost-cutting and sustainable targets in the textile industry [134].

4.1.4 Role of AI in Recycling Post-consumer Textiles

Post-consumer textile recycling has emerged as a key component in textile industry's circular economy. Despite this rising trend, only > 15% of textile waste is recycled globally and under 1% are reprocessed into new apparels [135]. The main reason is the complex mixture of fiber blends, dyes and non-textile constituents that make sorting and material recovery difficult. Recent AI automated innovations like Fibersort have significantly enhanced the ability of identifying and sorting textiles for recycling. Fibersort works on the principle of combining hyperspectral imaging with computer vision for precisely determining the composition of fabric up to 95% accuracy. The H&M Foundation and HKRITA's Smart Sorting System segregates garments based on fiber content and color within milliseconds, overtaking manual sorting methods in both speed and precision [136]. A U.S. firm called Refiberd was found in 2025 and developed a device that can recognize intricate fiber combinations like cotton and elastane. This invention prevents about 70% textile waste from entering landfills and enables more efficient textile-to-textile recycling [137]. Such advanced and enhanced recognition of material types makes methods of mechanical and chemical recycling more viable and sustainable. Further advancements include incorporation of AI in robotic assembly systems. These devices are being designed to automate processes that previously required manual labor, like removal of buttons, zippers and other non-recyclable parts [128]. By precisely determining material compositions and regulating treatment parameters, AI simultaneously enhances chemical recycling processes, thereby boosting recycling yields and reducing adverse environmental consequences [128]. AI-driven technologies not only reduce the reliance on textile waste export but also promote expansion of domestic recycling infrastructures [128].

4.2 Technological Advances in Textile Waste Sorting for Recycling Efficiency

Effective sorting of textile waste is very essential for efficient recycling, as it directly impacts the yield and quality of recovered fibers and end-products. Due to the wide range of fabric types, dyes, finishes and other pollutants present in the textile waste streams add complications and renders effective sorting difficult [128]. Novel approaches for overcoming these obstacles are made possible by recent technical advancements. Combining AI and hyperspectral imaging is one of the strategies that makes use of specialized cameras that can take photographs at different wavelengths. High level of precision and operating speed are achieved by processing these pictures using AI-driven algorithms that segregate textiles according to fiber type and presence of contaminants [138]. Another widely applied technique implies the use of near-infrared (NIR) spectroscopy combined optical sorting techniques. NIR spectroscopy works on the principle of measuring the wavelengths of near-infrared light reflected by different textile materials. Along with the detection of dyes and surface treatments, these reflected spectrum fingerprints, enabling separation of fiber types like cotton, silk, wool and polyester [139]. This spectral information is used by sorting machines to appropriately segregate textiles. PICVISA, a commercial application based on this same principle was developed in Barcelona. Large-scale automated sorting is achieved by using PICVISA, which integrates NIR spectroscopy with an RGB (red, green, blue) camera system to differentiate fabrics based on both color and material composition [140]. Similarly the WASTEX system effectively sorts textile wastes and produces high-quality raw material for the manufacturing of recycled yarn by combining AI technology with NIR and RGB [141]. These sorting technologies despite having potential, come along with certain drawbacks. High-performance computer system, sensitive optical detectors and large storage data capacities are required for implementation of AI algorithms to process the substantial amount of spectral data. Technical problems like inconsistent brightness and pixel-level abnormalities can also affect hyperspectral and NIR imaging systems, which can hinder algorithm performance and lead to errors in segregation [142]. Nevertheless the ongoing expansion and commercialization of sophisticated textile sorting technologies point to a path toward effective and economical recycling methods.

4.3 *Cotton Waste for Renewable Energy Recovery*

Cotton waste, particularly cotton gin waste (CGW) as well as cotton stalks, holds substantial potential as a substrate for renewable energy generation due to its high energy value, ranging between 18.6 and 20.9 MJ kg^{-1} [39]. Various conversion processes have been studied to convert this agricultural waste into useful forms of bioenergy, including pyrolysis, anaerobic digestion, bioethanol production and pelletization as shown in Fig. 2.

4.3.1 Pelletization of Cotton Waste Residues for Fuel Pellet Production

The process of pelletization is mainly affected by temperature and the lignin content of the cotton residues. The natural content of lignin is responsible for the binding of material in the pellets upon cooling [39]. It is proven that fuel pelletization or briquette formation from cotton waste residues have advantages like low transportation and maintenance costs and reduced dust emissions and thus can be a financially viable strategy. One drawback is the mechanical fragility of these pellets as these are prone to breakage and abrasion and can compromise the energy efficiency [143]. United States Department of Agriculture—Agricultural Research Service (USDA-ARS) developed a patent COBY (Cotton Byproduct) process for generating numerous value-added products from starch and other various additives before getting heat, pressure and blending treatment to form high-quality fuel pellets [143]. The briquette production cost from cotton stalks is approximately $89.26/ton, marginally very less than the

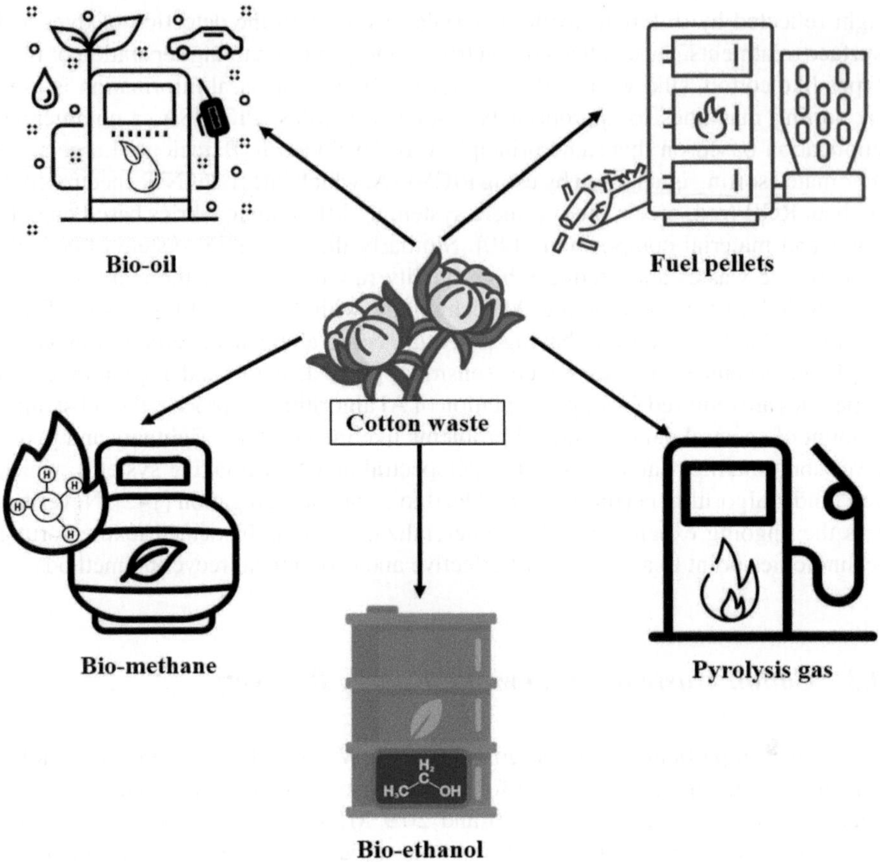

Fig. 2 Cotton waste for renewable energy recovery

market price of wood charcoal which is \$170–\$229/ton [143]. It was reported that cotton stalks have a calorific value of 20.9 MJ kg^{-1}, when carbonized and made briquettes [39].

4.3.2 Bioethanol Production from Cotton Waste

Due to their high lignocellulosic content (particularly, cellulose and hemicellulose) of 35–49%, cotton stalks and gin waste are a potential source of bioethanol production [144, 145]. Implying steam explosion, enzymatic hydrolysis and fermentation using *Escherichia coli* KO11 ethanol production was at 83% of the theoretical maximum. This further improved the hydrolysis efficiency from 42 to 67% [146]. Out of the different chemical pre-treatments tested to improve the saccharification of cotton stalks, sodium hydroxide treatment (2% NaOH for 90 min at 121 °C/103.4 kPa), attained 65.63% delignification and superior cellulose conversion in comparison with hydrogen peroxide, sulfuric acid and ozone treatments [145].

4.3.3 Bio-oil and Pyrolysis Gas

Another by-product extraction method, pyrolysis liquefaction process is used for converting cotton waste into valuable energy products like char, gas and bio-oil. This process involves thermochemical decomposition of cotton waste residues at very high temperatures (approximately 477 °C) in oxygen-deficient conditions [147, 148]. The resulting bio-oil is a mixture of various compounds such as methanol, acetic acid, phenolics and carbonyls [149]. This oil can be refined into civil gas or other chemical byproducts [150]. Out of the three treatments of $ZnCl_2$, Na_2CO_3, and K_2CO_3 at temperatures up to 751.9 °C, highest hydrogen-rich gas yield (27.0% by weight) was showed by Na_2CO_3, in the process of catalytic pyrolysis of cotton cocoon shell waste [151, 152]. In another study cotton stalks pyrolyzed at 850 °C in a fixed-bed reactor resulted in 58% gas, 20% water, 17% char and 5% tar. The gases were primarily CO and H_2 having a calorific value of 14,000 kJ Nm^{-3} [150].

4.3.4 Bio-methane Generation via Anaerobic Digestion

Through anaerobic digestion, cotton waste can be used for producing biogas, especially when co-digested with animal dung for balancing the carbon-to-nitrogen (C:N) ratio and alleviate toxicity of ammonia [153]. A two-stage digestion system using dairy manure to supply nutrients and cellulolytic micro-organisms produced a methane yield of 87 m^3 ton^{-1} of waste in a study by Funk et al. [154]. There was a 35% increase in digestion over digestion of dairy manure solely [155]. The digestate had twice the nitrogen content in comparison with other composted residues, thus making it valuable source for soil amendment [154]. Along with generation of renewable energy this approach provides a sustainable method for disposal of cotton waste [156].

5 Sustainable Approaches for Upcycling and Recycling Cotton Waste

In the year 2018, cotton alone contributed to one-third of the entire textile production and was further projected to raise by 6.5% in 2019 [45]. However, significant pressure is exerted on the environment by this expansion, particularly in terms of water consumption land usage and chemical inputs. In the year 2004, cotton cultivation was accountable for around 2.6% of the global water footprint [157]. Moreover, it is an extremely chemical-intensive crop, accounting for approximately 11% of the global pesticide usage till 2013 [158]. Traditionally, virgin cotton is widely exploited in the apparel industry, which characteristically goes for disposal post-consumer use. The U.S. Environmental Protection Agency [159] reportedly confirmed that annually about 9.5 million metric tons of textile as well as apparel waste is sent for disposal in landfills in the US alone, but hardly 1% of this waste is recycled into new garments [160]. However, it has been estimated that above 80% of cotton waste have the potential to be mechanically recycled for textile production [160]. Additionally, utilizing recycled cotton can potentially alleviate dependence on water, dyeing chemicals, pesticides and energy, thereby minimalizing environmental pollution and amount of waste [161]. Efforts are made to enhance the reuse and recycling of cotton waste across various sectors like policy, industry, and academia. In the policy domain, the EU introduced the Circular Economy Action Plan in 2015, a 3-year initiative to promote material reuse, with a particular focus on the textile sector [162]. In the US, the Council for Textile Recycling (CTR) recorded 40% rise in textile waste from 1998 to 2009, complemented by only a 2% improvement in waste deviation from landfills. Consequently, the Council has intended to achieve zero textile waste in landfills by the year 2037 [45]. In the industrial domain, several enterprises are working for large-scale mechanical and chemical recycling technologies. TeKiDe in Finland and Renewcell in Sweden are adopting chemical recycling of used cotton garments [163, 164]. A Spanish company, Recover, emphases on mechanical recycling by shredding pre-consumer cotton waste but still carefully preserving the staple length of fibers, allowing re-spinning [165]. Recognized companies are also incorporating recycled cotton into their supply chains. Lenzing, a prominent brand in renewable textile fibers, introduced Refibra technology in 2016, which utilizes pre-consumer cotton waste in combination with wood pulp. The future roadmap includes integrating post-consumer cotton waste as well [45]. Chief fashion brands like Levi's, H&M and The North Face have launched consumer involvement programs to collect used clothes. H&M, reported a rise in its textile collection bulk from 12,000 to 21,000 metric tons between 2015 and 2018 [166]. Academic research has worked parallel with industrial sector by tackling basic technical barriers accompanying with cotton recycling. These include upgraded characterization of mixed textile waste types [167], enhancing the effectiveness of segregating cotton from polyester blends [168, 169], and creating value-added products from cotton waste that would probably have the potential to substitute virgin materials [170]. Various reuse and recycling methods have been investigated, which can be roughly classified into 5 main categories: reuse, fabric recycling, fiber recycling, polymer or monomer recycling (primarily for synthetic materials or blends), and energy recovery [171, 172]. Recycling strategies such as fabric and fiber recycling give rise to either downcycled or upcycled products, each with different economic significance and implications for different stakeholders throughout the textile value chain [172]. Reuse and recycling methods, predominantly cloth reuse and fabric repurposing, have demonstrated potential in plummeting environmental burdens by compensating the need for virgin textile production [61, 173, 174]. Nevertheless, this environmental benefit may involuntarily suppress demand for new textile materials. In situations where recycling is impractical, redesigning the product with recyclability in mind has emerged as a viable alternative. This is often referred to as "design for recycling" [175, 176]. Cotton recycling methods primarily fall into two categories: mechanical and chemical recycling.

5.1 Mechanical Recycling

Mechanical recycling initiates with sorting and cleaning cotton fabrics, followed by shredding them into fibers which are suitable for re-spinning into yarn or transformation into nonwoven materials [177]. Conversely, the process reduces fiber length and deteriorates quality, particularly in terms of strength and softness. Therefore, post-consumer waste which often results in lower fiber quality is not preferred, instead pre-consumer waste is typically favoured [45]. The integration of mechanically recycled fibers in fresh fabric is limited to 20–30% for avoiding significant quality loss [45]. Some technological advancements like "Recover" have improved mechanical recycling outcomes. Recover, has developed a method that conserves fiber staple lengths up to 15 mm (in comparison to the usual < 10 mm) thus enabling increase in the recycled content in blends [165]. LCAs, by Miljögiraff for H&M, demonstrated that mechanically recycled cotton overtakes virgin fiber in numerous environmental metrics. In this LCA, quality degradation was not assessed [161]. An evaluation was conducted on mechanically recycled textile waste material and it was found that such products when used in oil absorbent mats, produced five times lesser GHG emissions than polypropylene mats. However, economic and quality assessments were not considered [178]. Further investigations on cotton-polyester blends demonstrated that electricity consumption in yarn spinning process significantly influenced the overall environmental footprint. When recycled cotton was blended with PET fibers, the environmental burden was largely influenced by upstream PET production [45]. Apart from yarn and fabric, mechanically recycled cotton fibers have been explored for utilization in composites, especially in automotive and construction sectors like [179]. Additionally, the potential of integrating cotton fibers into cement-based mixtures for construction when explored, revealed promising mechanical performance [180, 181]. Solvaira Specialties a US-based company, has also demonstrated the efficacy of micronized recycled cotton from used T-shirts (processed to dimensions of 350 μm in length and 20–40 μm in width) in applications like composites, 3D printing materials and nanocellulose production [165]. Additional applications include its use in paper, filtration devices, and hygiene products, reflecting the broad potential of cotton waste in diversified product domains.

5.2 Chemical Recycling

Chemical recycling of cotton waste is evolving as a dynamic strategy to address environmental concerns related to textile waste and to encourage circular economy. The characteristic composition of cotton, primarily cellulose, enables its conversion into valuable bio-based products and chemicals via thermochemical and biochemical approaches. Sandin and Peters [174] stressed the necessity for shifting from linear to circular textile systems, emphasizing the environmental benefits of fiber-to-fiber recycling technologies. Innovations like REFIBRATM, incorporate post-consumer cotton waste into the production of Lyocell fibers, giving a closed-loop recycling model [45, 167]. When the potential of cotton and other lignocellulosic materials for biopolymer production was explored, it indicated that chemical treatments can boost cellulose recovery, which is essential for downstream processing into nanocellulose and several other advanced materials. The valorization of cotton waste through fermentation and anaerobic digestion has gained attention recently. Hamawand et al. [39] demonstrated the bioenergy potential of cotton lignocellulosic biomass, encouraging its incorporation in renewable energy structures. One of the studies implemented enzymatic pre-treatment of lignocellulosic biomass which resulted in efficient hydrolysis of cellulose into fermentable sugars, which is a critical step in bioethanol production [182]. Biochar production from cotton waste biomass via pyrolysis, demonstrates its applications in agriculture and water treatment [45]. Under optimized conditions enzymatic hydrolysis of pretreated cotton fibers resulted in substantial yields of bioethanol [183]. A widespread review of nanocellulose derived from cotton and various lignocellulosic sources, emphasized its role in polymer syntheses and biomedical applications [184]. In terms of waste management strategies, a classification system for textile end-of-life treatments was proposed, with chemical recycling being a key component in closing material loops [171]. Pyrolysis and combustion techniques, are found to be beneficial in both energy recovery and waste reduction [45]. The processes of converting cotton waste into biofuels has been extensively studied and documented. Various pre-treatment techniques are studies and demonstrated for improving the lignocellulosic digestion of agriculture biomass [185]. Ranjithkumar et al. [186] optimized microbial fermentation for enhanced bioethanol production. The sustainability of cotton-based substrates in anaerobic digestion systems was confirmed after yielding methane-rich biogas [187, 188]. Sharma-Shivappa and Chen [189] compared enzymatic and dilute acid hydrolysis methods, discovering that process efficacy profoundly depends on the characteristics of feedstock. Cotton waste as a potential input was proven in the evaluation of biogas yields from various textile wastes. This is majorly due to its high cellulose content [41]. The environmental advantages of cotton-derived biochar and biochar in general, including pollution control, carbon sequestration and soil remediation, have been emphasized by Gwenzi et al. [190]. Barrow [191] discussed the socio-ecological effects of bio-based systems, emphasizing the significance of sustainable waste utilization. Rago et al. [192] demonstrated successful incorporation of cotton waste into thermochemical valorization methods such as gasification and pyrolysis. Hybrid catalytic systems for converting cellulose into platform chemical compounds are being proposed as a novel and an innovative approach to cotton waste valorization [193]. The surface modification of regenerated cellulose from cotton for application in supercapacitors and adsorbents was investigated, indicating a promising method for electronic and environmental applications [194, 195]. Cotton-derived cellulose was utilized for developing water treatment membranes using, demonstrating the diverse applications of recycled cotton [196]. Activated carbon derived from textile waste was utilized and applied for effective adsorption of heavy metal adsorption from contaminated water [197]. A particular LCA for cotton waste valorization strategies, confirmed the environmental superiority of chemical recycling in comparison with incineration and landfilling [198]. There is rising emphasis on the importance of infrastructure and policy that supports cotton recycling establishing a link between technology readiness and adoption of circular economy [199]. Advanced gasification techniques relevant to cotton biomass when explored, highlighted the process effectiveness and potential of energy recovery [200]. Table 3 gives a comparative overview of both mechanical and chemical recycling of cotton textiles.

6 Integrating Textile and Cotton Waste into the Circular Economy Through Waste Valorization

6.1 Shift of Textile Wastes to Circular Economy

Historically, the resources consumed by living organisms were naturally reincorporated into the environment for millions of years before the industrial development. However, the advancement of material technologies, which have become more complicated and challenging to recycle has shifted caused modern society to move away from this sustainable cycle. Waste management has become a serious responsibility of the modern society since it is not a natural concept but rather a product of human consumption and production activities [201]. In past few decades, waste management (WM) has become a major global concern, with the textile sector being the highest contributor to this issue [202]. Textile production and clothing consumption have become essential to daily life and play a vital role in global economic systems. Despite these contributions, textile industry is widely known for its harmful and detrimental environment and health effects which include overuse of chemicals and water as well as rising textile waste which is estimated to roughly 20% of total global waste [17, 203, 204]. Currently, the textile sector predominantly follows a linear economic model, characterized by a one-way flow of resources, beginning with raw material extraction, manufacturing, distribution, consumption, and ends with disposal often in landfills or undergoes incineration. This destructive approach causes enormous losses in terms of both resources and energy [205]. It is estimated that less than 10% of the global economy follows circular principles, emphasizing the scale of waste and inefficiency levels [206]. As a result, there is a growing need for innovative technologies that make it possible to recover and repurpose valuable materials from textile waste streams. There is increasing recognition of textile waste as a treasured secondary resource. However, a lack of suitable technology and policy frameworks this potential is often underutilized. Implementing a circular economy approach in the textile industry, has been gradually promoted as a solution to alleviate the negative social and environmental impacts of fashion industry [174]. In

Table 3 Comparison between mechanical and chemical recycling

Parameters	Mechanical recycling	Chemical recycling	References
Initial process	Sorting, cleaning, shredding into fibers	Pre-treatment (e.g., enzymatic, acid, or ionic liquid treatment), followed by depolymerization	[177, 182]
Material output	Shortened fibers for yarn/ nonwoven, composites, insulation	Cellulose derivatives, bioethanol, nanocellulose, biogas, biochar, chemicals	[180, 184, 188]
Fiber quality	Reduced length (< 10 mm), strength, softness	Retains or transforms cellulose into pure forms	Liu et al. (2019)
Preferred waste type	Pre-consumer	Both pre- and post-consumer	[45]
Recycled content limit in fabric	20–30% to avoid quality loss	Can be completely regenerated (e.g., into Lyocell fibers)	[45]
Environmental impact (LCA)	Lower GHG emissions in applications like oil absorbent mats; better than virgin cotton in many metrics	Environmentally superior to incineration and landfill; enables circular economy	[161, 178, 198]
Energy considerations	Yarn spinning consumes significant electricity; affected by PET blending	Energy recovered via biofuels, gasification, pyrolysis	[45, 192, 200]
Applications	Yarn, fabric blends, automotive and construction composites, 3D printing, paper, hygiene products	Bioethanol, biogas, nanocellulose, supercapacitors, water treatment membranes, platform chemicals	[165, 194, 196]
Technology examples	Recover (fiber length up to 15 mm), Solvaira (micronized cotton)	REFIBRATM, enzymatic hydrolysis, ionic liquid treatment, hybrid catalytic systems	[45, 165, 193]
Challenges	Limited fiber quality, economic feasibility, blending limits	Technological readiness, infrastructure and policy support needed	[45, 199]

addition to sustainable consumption and production models, the circular economy emphasizes resource efficiency through various practices like reuse, recycling, repair, and refurbishing [203]. Biological treatment strategies, particularly composting and anaerobic digestion are amongst the most potential methods which align with circular economy principles. These methods can transform textile waste into valuable products like biofertilizers and biogas. These methods provide a sustainable alternative to conventional mechanical or chemical recycling, which often deteriorates fiber quality and increases pollution [207]. These biological methods help renewable nutrients recovery, improve fertility of soil, and enable generation of clean energy. Moreover, global textile production is estimated to exceed 99 million tons annually [202], and will further increase by 60% between 2015 and 2030, reaching 148 million tons per year. 73% of this waste is either incinerated or landfilled which incurs a loss of more than $100 billion in raw material value [208]. An estimated $192 billion could potentially be gained by 2030, if the textile industry adopts sustainable methods of operation [209].

6.2 Advancements in Textile Waste Management to Enhance Circularity in Textiles

Because there is no infrastructure to manage this enormous amount of textile waste generated, particularly in urban areas, the textile waste management has become increasingly challenging [127]. Textile waste can be recycled into value-added products instead of being thrown away, which is why design and production strategies are shifting towards more sustainable, reusable and recyclable methods [210]. This transition aligns with international government policies to encourage recycling minimize landfilling. Even though incineration poses a lower risk to human health (28% of disease outcomes) compared to landfilling (46.8% of disease outcomes) both are unsustainable and impose negative effects on the ecosystem [211]. This emphasizes the need for cleaner and more sustainable alternatives. In this regard, a substantial amount of research has explored methods to recover, reuse, and recycle textile waste that are both cost-effective and sustainable, closely adhering to the circular economy principles [212]. The circular model aims to extend material life cycles, improve recycling methods, and optimize material usefulness, thereby reducing resource consumption and overall impact on the environment [211]. It products are aimed to send back into their same production value chain [213]. The 5R framework of reduce, refuse, reuse, recycle, and recover, has gained international recognition as the foundation of sustainable waste management [211]. Sustainable methods in textile production include alternatives like waterless dyeing and reducing water use in finishing processes [214, 215]. Environmental concerns have deepened due to the increasing accumulation of non-biodegradable textile waste and unavailability of landfills, stimulating the adoption of environmental-friendly strategies across the entire textile product life cycle [208]. According to some studies it is stated that 95% of textile waste is recyclable and can be transformed into useful products [202], with approximately 60% of recovered garments suitable for reuse and 35% for conversion into fiber materials or rags. Waste reduction strategies like reducing overproduction and designing for durability and recyclability are essential to solve these issues. Sorting and segregating by color and type is usually the first step in recycling natural fibers, since it minimizes the use dyeing, conserves energy and water as well as alleviates pollution level [216]. Mechanical, chemical, thermal, and hybrid recycling methods are generally used, and their classification usually depends on the input waste and output product. According to estimates, 50% of denim products and 30% of new garments are made from recycled cotton, occasionally in the form of blended yarns with synthetic fibers [217]. Separation processes are required for separation of components like cotton and PET from blended textiles like polycotton. Chemically this can be achieved by depolymerizing one component while maintaining the other, although such methods can have environmental and health risks [174, 218]. Another environmental concern in addition to textile recycling is sound pollution. Conventional soundproofing materials like mineral wool or foams made from petrochemicals are not sustainable because they consume very high energy, less degradable and pose several health risks [219]. As a result, alternative applications for recycled textile waste have developed, particularly in insulation. Research has advanced the development of biocomposites for building insulation that use natural and recycled fibers [220], providing an economical and resource-efficient alternative to conventional materials. In addition mitigating the potential environmental effects, recycled insulating materials have useful benefits for various sectors like construction and auto-upholstery [203]. Textile waste is a potential component of green building materials owing to its physical properties which makes it suitable for thermal and acoustic insulation [215]. Utilizing these secondary fibers results in reduction of the need for virgin resources, thereby lowering greenhouse gas emissions, reducing the impact on ecosystems and improving public health [221]. By transforming otherwise discarded fibers into inexpensive insulation, recycling encourages new market prospects and have a potential to boost economy. Reliance on non-renewable sources is reduced when insulation is made from natural, renewable sources [219]. Despite technological development, textile recycling remains limited, main reason being quality degradation with repeated cycles. Given that most existing methods result in downcycling, there is an evident need for innovative and sustainable alternatives. Hence, biological treatments are being frequently employed as a sustainable method for future textile waste processing [127].

6.3 Sustainable Cotton Waste Valorization and Its Applications

Cotton waste has been emerging as a valued resource for several industrial and environmental applications. Its lignocellulosic content, abundance, and biodegradable properties bring its potential for reuse in multiple areas.

6.3.1 Sustainable Adsorbent from Cotton Residues

Activated carbon is normally used to remove pollutants from water and air due to its large surface area [222]. Particularly when these adsorbents are produced from agricultural residues like cotton waste, it has several advantages like cost-effectiveness, mechanical robustness, and effective adsorption capacity [223]. Using steam and zinc chloride activation, Attia et al. [224] investigated the adsorption properties of activated carbon made from cotton stalks. Significant surface area porosity and dye-takeup ability were displayed by the final product. Additionally, it was discovered that textile dye removal from wastewater may be achieved with activated carbon made from silk cotton hulls treated with concentrated sulfuric acid [225]. According to Kadirvelu et al. [226], this carbon variant has a removal efficiency of 89% and can instantly remove dyes including Congo red, methyl violet, malachite green as well as mercury (II) ions. According to Hashem et al. [227] acid dyes can be effectively removed from water using quaternized cotton stalks modified by grinding and treating with Refaktam K. Furthermore, untreated cotton waste, after cleaning, drying, and shredding, has displayed high adsorption capabilities for methylene blue and Safranine dyes, with saturation values of 277 and 190 mg g^{-1} respectively [228].

6.3.2 Cotton Residues as an Alternative in Animal Feed

Cotton residues can be used as animal feeds due to their high cellulose content [189]. Components like stalks, cottonseed hulls, cotton gin waste (CGW), linters, and motes have been used in winter feeds to cut feeding costs [229]. A study by the University of Georgia proved that cattle given cotton stalks gained weight comparable to those cattle who were fed coastal bermudagrass hay, demonstrating nutritional potential of cotton stalks [229]. Although CGW has potential utility, its poor digestibility in raw form has resulted in the application of chemical treatments particularly ammonium hydroxide and sodium hydroxide to enhance its nutritional properties and enhance digestion [189]. It was reported that treating cotton gin trash (CGT) with 6% NaOH at room temperature (RT) enhanced its digestibility to 70.5% [230].

6.3.3 Substrate Cultivation Bed for Mushrooms

Cotton residues being lignocellulosic wastes are efficient, low-cost substrates for mushroom cultivation, offering a cotton waste recycling method [189]. These wastes, composed of cellulose, hemicellulose, and lignin, undergo degradation by fungi through enzymatic action [189]. Philippoussis et al. [231] used Cotton stem-leaf residues and CGT were used to cultivate mushroom species like *Pleurotus eryngii*, *P. ostreatus*, *P. pulmonarius*, *Volvariella volvacea*, and *Agrocybe aegerita*. High colonization, rapid fructification, and substantial yields were reported. It was also found that *Volvariella esculenta* began fruiting 13 days post-inoculation on cotton waste residues, which indicated that cotton residues outperformed other agricultural residues in yield mainly due to its higher cellulose content as well as compact structure. After harvesting, the remaining cotton stalks can also be processed into pulp and pulp production efficiency can be improved by pretreating stalks with 1% NaOH followed by anaerobic fermentation [189].

6.3.4 Enzyme Synthesis Using Cotton Waste as a Substrate

In solid-state fermentation systems, cotton waste can also be used as an inexpensive substrate for the manufacture of enzymes [189]. It is an inducer as well as a source of carbon. The cellulase enzyme complex, which includes β-glucosidase, endo-1, 4-D-glycanase, and exo-1,4-D-glucanase, is inducible in such conditions [232]. *Pleurotus sajor-caju* cultivated on alkali-treated cotton waste had higher activity of enzymes such as cellobiohydrolase (15.6 U mg^{-1} protein), CMCase (83.4 U mg^{-1} protein), and β-glucosidase (56.1 U mg^{-1} protein) than sawdust [189]. Cotton leaf residue was found to be a practical and cost-effective substitute for peptone in the synthesis of xylanase, despite the fact that the enzyme yield (22.1 U mg^{-1} protein) was less than that of peptone (44.9 U mg^{-1} protein) [233].

6.3.5 Cotton Waste for Composting and Soil Management

Cotton waste is highly beneficial and vital in composting, serving as a structural agent to control moisture and enhance porosity [234]. Aeration is obligatory for efficient composting in order to promote microbial activity and allow heat and moisture to dissipate [189, 235, 236]. For composting with high-moisture organic waste like swine manure and olive mill byproducts, low-moisture agricultural residue like CGW are the perfect bulking agents [189]. Cotton waste increases cation exchange capacity (CEC), decreases phytotoxicity, and enhances C:N ratios in compost [189, 237]. According to Paredes et al. [234], composting olive mill sludge with cotton waste produced higher-quality compost than utilizing maize straw. Because of its fibrous nature, COBY, a cotton-based mulch created by USDA-ARS, demonstrated exceptional soil adhesion and erosion prevention [238, 239].

6.3.6 Cotton Stalks for Making Particleboards in Construction

Particleboards can be made from chipped cotton stalks by hot pressing them with hardeners like ammonium chloride and adhesives like urea formaldehyde. Guler and Ozen [240] discovered that boards produced at 150 °C for 6 min under pressure (2.4–2.6 N mm^{-2}) satisfied industrial performance requirements, such as structural integrity and water resistance. In order to avoid contaminating groundwater, cotton waste has also been investigated as a natural sealant for anaerobic lagoons. It was shown that compacting cotton gin waste with gravel and dirt greatly decreased hydraulic conductivity (2.2×10^{-6} cm s^{-1}), suggesting that it could be used as a biological liner [189].

7 Conclusion

Due to the rising global environmental problems, understanding and management of carbon footprints has become very crucial for individuals, companies, and countries as a whole. This study has focused on the environmental threats posed by cotton and textile industries associating with the concept of carbon footprint as a quantitative assessment of GHG emissions, predominantly carbon dioxide, released directly or indirectly through anthropogenic activities. Although the cotton-based textile and fashion industries play a critical role in development of global economy, they are also amongst the leading contributors to widespread carbon emissions and environmental damage. The carbon footprint associated with cotton cultivation and production-processing through textiles has been comprehensively studied in this study, demonstrating the various challenges faced by agricultural and industrial sectors in sustainable cotton waste management. Despite its economic importance, cotton cultivation and production is highly resource-intensive, consuming substantial amount of water and agrochemicals. Also GHG emissions, primarily from fuel consumption and fertilizer use, leads to a huge carbon footprint of cotton on the environment. Industrial processes like spinning, dyeing, and apparel manufacturing further exacerbate this impact because of high energy requirements and substantial amount of waste generation. Moreover, the increasing trend of fast fashion has caused an exponential growth in both pre- and post-consumer textile waste, and a large part of this waste is either incinerated or ends up in landfills further increasing the carbon footprint. The study emphasizes the potential as well as importance of recycling strategies and sustainable waste management for mitigating these impacts. Energy recovery processes such as anaerobic digestion and pyrolysis along with mechanical and chemical recycling methods, have tremendous potential in converting cotton waste into value-added resources like biogas, bioethanol and biochar. Artificial intelligence is being increasingly used and integrated into waste sorting and textile manufacturing which further reduces waste, conserves water and energy and improves material efficiency. Transitioning and moving toward circular economy, in which cotton waste is considered as a valuable secondary resource instead of being thrown away, emerges as an innovative and sustainable solution. This involves systemic changes in production, consumption as well as disposal practices, sustained by technological advancements, policy frameworks, and consumer involvement. In conclusion, a multi-faceted and extensive approach is necessary for reducing the carbon footprint of the cotton textile industry. This approach will span the entire cotton textile value chain from sustainable cultivation and efficient manufacturing to cutting-edge recycling methods and waste valorization. A practical blueprint for combating severe environmental impacts while increasing resource efficiency is possible by the combination of AI-powered technologies, renewable energy integration, and circular economy principles provides. Such holistic strategies not only address climate change but also encourage resilience and sustainability in the global textile sector.

References

1. United Nations Climate Change (2018) UN helps fashion industry shift to low carbon. UNFCCC. https://unfccc.int/news/un-helps-fashion-industry-shift-to-low-carbon
2. The Carbon Trust (2011) International carbon flows. Clothing. CTC793. The Carbon Trust. https://prod-drupalfiles.storage.googleapis.com/documents/resource/public/International%20Carbon%20Flows%20-%20Clothing%20-%20REPORT.pdf
3. Quantis (2018) Measuring fashion: insights from the environmental impact of the global apparel and footwear industries. Full report and methodological considerations. Quantis. https://quantis-intl.com/measuring-fashion-report
4. Global Fashion Agenda (GFA) & The Boston Consulting Group (BCG) (2017) Pulse of the fashion industry. globalfashionagenda.com https://www.globalfashionagenda.com/wp-content/uploads/2017/05/Pulse-of-the-Fashion-Industry_2017.pdf
5. Dahlbo H, Aalto K, Eskelinen H, Salmenperä H (2017) Increasing textile circulation—consequences and requirements. Sustain Prod Consump 9:44–57
6. Ellen MacArthur Foundation (2017) A new textiles economy: redesigning fashion's future. https://ellenmacarthurfoundation.org/a-new-textiles-economy
7. Peters GM, Sandin G, Spak B (2019) Environmental prospects for mixed textile recycling in Sweden. ACS Sustain Chem Eng 7(13):11682–11690
8. Fletcher K (2016) Craft of use: post-growth fashion. Routledge
9. Remy N, Speelman E, Swartz S (2016) Style that's sustainable: a new fast-fashion formula. McKinsey & Company. https://www.mckinsey.com/business-functions/sustainability/our-insights/style-thats-sustainable-a-new-fast-fashion-formula
10. Anguelov N (2015) The dirty side of the garment industry: fast fashion and its negative impact on environment and society. CRC Press
11. Niinimäki K (2018a) Eco-friendly and fair: fast fashion and consumer behaviour. In: Becker-Leifhold C, Heuer M (eds). Routledge, pp 49–57
12. Niinimäki K (2011) From disposable to sustainable: the complex interplay between design and consumption of textiles and clothing. Doctoral dissertation, Aalto University
13. Gardetti MA, Torres AL (eds) (2017) Sustainability in fashion and textiles: values, design, production and consumption. Routledge
14. EEA (European Environment Agency) (2014) Environmental indicator report: environmental impacts of production–consumption systems in Europe. https://www.eea.europa.eu/publications/environmental-indicator-report-2014
15. Laitala K, Klepp IG (2015) Product lifetimes and the environment. In: Cooper T et al (ed) Proceedings of the product lifetimes and the environment conference. Nottingham Trent University, pp 182–186. http://www.plateconference.org/pdf/plate_2015_proceedings.pdf
16. Petter O (2019) Brits to spend £2.7 bn on outfits they wear once this summer. The Independent, 20 May 2019. https://www.independent.co.uk/life-style/fashion/summer-outfits-spend-billions-fast-fashion-barnardos-charity-shop-a8998846.html
17. Niinimäki K, Peters G, Dahlbo H, Perry P, Rissanen T, Gwilt A (2020) The environmental price of fast fashion. Nat Rev Earth Environ 1(4):189–200
18. Turker D, Altuntas C (2014) Sustainable supply chain management in the fast fashion industry: an analysis of corporate reports. Eur Manag J 32(5):837–849
19. Munir H, Rasul F, Ahmad A, Sajid M, Ayub S, Arif M, Khan MA (2020) Diverse uses of cotton: from products to byproducts. In: Cotton production and uses: agronomy, crop protection, and postharvest technologies, pp 629–641
20. Ashraf S, Ch KM, Ashraf I, Akbar N (2024) A phenomenological inquiry into farmers' experiences growing cotton in Punjab, Pakistan. Sci Rep 14(1):13846
21. Jabran K, Chauhan BS (eds) (2019) Cotton production. Wiley
22. Peters G, Li M, Lenzen M (2021) The need to decelerate fast fashion in a hot climate—a global sustainability perspective on the garment industry. J Clean Prod 295:126390
23. Ranasinghe L, Jayasooriya VM (2021) Ecolabelling in textile industry: a review. Resour Environ Sustain 6:100037

24. Sadowski M, Perkins L, McGarvey E (2021) Roadmap to net zero: delivering science-based targets in the apparel sector. World Resources Institute, pp 1–40
25. Leal Filho W, Perry P, Heim H, Dinis MAP, Moda H, Ebhuoma E, Paço A (2022) An overview of the contribution of the textiles sector to climate change. Front Environ Sci 10:973102
26. Leal Filho W, Dinis MAP, Liakh O, Paço A, Dennis K, Shollo F, Sidsaph H (2024) Reducing the carbon footprint of the textile sector: an overview of impacts and solutions. Text Res J 94(15–16):1798–1814
27. OECD/FAO (2024) OECD-FAO agricultural outlook 2024–2033. OECD Publishing
28. Yang Y (2019) A unified framework of life cycle assessment. Int J Life Cycle Assess 24:620–626
29. Huang W, Wu F, Han W, Li Q, Han Y, Wang G, Wang Z (2022) Carbon footprint of cotton production in China: composition, spatiotemporal changes and driving factors. Sci Total Environ 821:153407
30. PE International (2014) Life cycle assessment (LCA) of cotton made in Africa (CmiA) https://cottonmadeinafrica.org/wp-content/uploads/Life-Cycle-Assessment-CmiA-2014.pdf
31. Jewell J (2017) LCA update of cotton fiber and fabric life cycle inventory. Cotton Today. https://cottontoday.cottoninc.com/wp-content/uploads/2019/11/2016-LCA-Full-Report-Update.pdf
32. Baydar G, Ciliz N, Mammadov A (2015) Life cycle assessment of cotton textile products in Turkey. Resour Conserv Recycl 104:213–223
33. Zhao M, Zhou Y, Meng J, Zheng H, Cai Y, Shan Y, Yang Z (2021) Virtual carbon and water flows embodied in global fashion trade—a case study of denim products. J Clean Prod 303:127080
34. Hedayati M, Brock PM, Nachimuthu G, Schwenke G (2019) Farm-level strategies to reduce the life cycle greenhouse gas emissions of cotton production: an Australian perspective. J Clean Prod 212:974–985
35. Avadí A, Marcin M, Biard Y, Renou A, Gourlot JP, Basset-Mens C (2020) Life cycle assessment of organic and conventional non-Bt cotton products from Mali. Int J Life Cycle Assess 25:678–697
36. Singh P, Singh G, Sodhi GPS (2022) Data envelopment analysis based optimization for improving net ecosystem carbon and energy budget in cotton (*Gossypium hirsutum* L.) cultivation: methods and a case study of north-western India. Environ Dev Sustain 24(2):2079–2119
37. Gemtos TA, Tsiricoglou T (1999) Harvesting of cotton residue for energy production. Biomass Bioenergy 16(1):51–59
38. Coates W (2000) Using cotton plant residue to produce briquettes. Biomass Bioenergy 18(3):201–208
39. Hamawand I, Sandell G, Pittaway P, Chakrabarty S, Yusaf T, Chen G, Hopf J (2016) Bioenergy from cotton industry wastes: a review and potential. Renew Sustain Energy Rev 66:435–448
40. Sun Z, Jin BS, Zhang MY, Liu RP (2008) Cotton stalk combustion in a circulating fluidized bed. Chem Eng Technol 31(11):1605–1614
41. Isci ASLI, Demirer GN (2007) Biogas production potential from cotton wastes. Renew Energy 32(5):750–757
42. Tunç Ö, Tanacı H, Aksu Z (2009) Potential use of cotton plant wastes for the removal of Remazol Black B reactive dye. J Hazard Mater 163(1):187–198
43. Halimi MT, Hassen MB, Sakli F (2008) Cotton waste recycling: quantitative and qualitative assessment. Resour Conserv Recycl 52(5):785–791
44. Dipali B, Raichurkar PP, Bandopadhyay SP, Akade A, Pachhade PM (2011) To study effect of comber noil percentage on yarn quality. Fibre2Fashion. https://www.fibre2fashion.com/industry-article/6023/to-study-effect-of-comber-noil-percentage-on-yarn-quality
45. Johnson S, Echeverria D, Venditti R, Jameel H, Yao Y (2020) Supply chain of waste cotton recycling and reuse: a review. AATCC J Res 7(1):19–31
46. Kasemset C, Chernsupornchai J, Pala-ud W (2015) Application of MFCA in waste reduction: case study on a small textile factory in Thailand. J Clean Prod 108:1342–1351

47. Fletcher K, Tham M (eds) (2015) Routledge handbook of sustainability and fashion. Routledge, p 15
48. Runnel A, Raiban K, Castel N, Oja D, Bhuiya H (2017) Creating a digitally enhanced circular economy. Reverse Resour
49. Niinimäki K (2018b) Sustainable fashion in a circular economy. Aalto University
50. Mathews B (2016) One third of all clothing "never sold". Ecotextile News, 21 Sept 2016, p 21
51. Paton E (2018) H&M, a fashion giant, has a problem: $4.3 billion in unsold clothes. The New York Times, 27(03)
52. Starn J (2017) Swedish power plant ditches coal to burn H&M clothes instead. The Independent, 17 Nov 2017. https://www.independent.co.uk/news/business/news/sweden-power-plant-hm-coal-burn-vasteras-stockholm-oil-discarded-products-a8073346.html
53. Hendriksz V (2017) H&M accused of burning 12 tonnes of new, unsold clothing. Fashion United. https://fashionunited.com
54. Reints R (2018) Burberry burned $37 million worth of products to protect its brand. Fortune. https://fortune.com/2018/07/19/burberry-burns-millions
55. Schrieber J (2017) Recycling fashion's remnants: residential and commercial textile waste. Des J. https://medium.com/design-journal/recycling-fashionsremnants-residential-and-commercial-textile-waste-e58a5f0c91a8
56. Lieber C (2018) Why fashion brands destroy billions' worth of their own merchandise every year. The New York Times, 7 Aug 2018. https://www.nytimes.com/2018/08/07/business/fashion-merchandise-destroyed.html
57. Kirchain R, Olivetti E, Miller TR, Greene S (2015) Sustainable apparel materials. Materials Systems Laboratory, Massachusetts Institute of Technology, pp 1–34
58. Daystar J, Chapman LL, Moore MM, Pires ST, Golden J (2019) Quantifying apparel consumer use behavior in six countries: addressing a data need in life cycle assessment modeling. J Text Appar Technol Manag 11(1)
59. Nørup N, Pihl K, Damgaard A, Scheutz C (2019) Quantity and quality of clothing and household textiles in the Danish household waste. Waste Manage 87:454–463
60. Allwood JM, Laursen SE, Rodríguez CM, Bocken NMP (2006) Well dressed? The present and future sustainability of clothing and textiles in the United Kingdom. University of Cambridge, Institute for Manufacturing. https://www.ifm.eng.cam.ac.uk/uploads/Resources/Other_Reports/UK_textiles.pdf
61. Woolridge AC, Ward GD, Phillips PS, Collins M, Gandy S (2006) Life cycle assessment for reuse/recycling of donated waste textiles compared to use of virgin material: an UK energy saving perspective. Resour Conserv Recycl 46(1):94–103
62. Koligkioni A, Parajuly K, Sørensen BL, Cimpan C (2018) Environmental assessment of end-of-life textiles in Denmark. Procedia CIRP 69:962–967
63. Mui YQ (2010) Consumer complaints have companies rethinking how to dispose of unsold inventory. The Washington Post, p A8
64. Maxwell D, McAndrew L, Ryan J (2015) State of the apparel sector report—water. A report for the global leadership award in sustainable apparel
65. Boucher J, Friot D (2017) Primary microplastics in the oceans: a global evaluation of sources, vol 10. International Union for Conservation of Nature (IUCN)
66. Gervet B (2007) The use of crude oil in plastic making contributes to global warming. Lulea University of Technology, Lulea
67. Muthu S (2014a) Assessing the environmental impact of textiles and the clothing supply chain. Springer, pp 1–194
68. Radhakrishnan S (2017) Sustainable cotton production. In: Sustainable fibres and textiles. Woodhead Publishing, pp 21–67
69. Evett SRB, Howell RL, Ibragimov TA, Hunsaker NM, Jensen DJ (2012) Crop yield response to water: the AquaCrop simulation model. FAO irrigation and drainage paper No. 66. FAO Press

70. Huang B, Zhao J, Geng Y, Tian Y, Jiang P (2017) Energy-related GHG emissions of the textile industry in China. Resour Conserv Recycl 119:69–77
71. Rosas F (2012) Fertilizer use by crop at the country level (1990–2010). Food and Agriculture Organization of the United Nations
72. EJF (Environmental Justice Foundation) (2007) The deadly chemicals in cotton. Environmental Justice Foundation in collaboration with Pesticide Action Network UK. London, UK. ISBN: 1-904523-10-2
73. Dey KR, Choudhury P, Dutta BK (2013) Impact of pesticide use on the health of farmers: a study in Barak Valley, Assam (India). J Environ Chem Ecotoxicol 5(10):269–277
74. Maggi F, Tang FH, la Cecilia D, McBratney A (2019) PEST-CHEMGRIDS, global gridded maps of the top 20 crop-specific pesticide application rates from 2015 to 2025. Sci Data 6(1):170
75. Zhang Z, Huang J, Yao Y, Peters G, Macdonald B, La Rosa AD, Scherer L (2023) Environmental impacts of cotton and opportunities for improvement. Nat Rev Earth Environ 4(10):703–715
76. Coupe RH, Capel PD (2016) Trends in pesticide use on soybean, corn and cotton since the introduction of major genetically modified crops in the United States. Pest Manag Sci 72(5):1013–1022
77. Kouser S, Spielman DJ, Qaim M (2019) Transgenic cotton and farmers' health in Pakistan. PLoS ONE 14(10):e0222617
78. Qaim M, De Janvry A (2005) Bt cotton and pesticide use in Argentina: economic and environmental effects. Environ Dev Econ 10(2):179–200
79. Maggi F, Tang FH (2021) Estimated decline in global earthworm population size caused by pesticide residue in soil. Soil Secur 5:100014
80. Bakhsh K, Ahmad N, Kamran MA, Hassan S, Abbas Q, Saeed R, Hashmi MS (2016) Occupational hazards and health cost of women cotton pickers in Pakistani Punjab. BMC Public Health 16:1–11
81. Memon QUA, Wagan SA, Chunyu D, Shuangxi X, Jingdong L, Damalas CA (2019) Health problems from pesticide exposure and personal protective measures among women cotton workers in southern Pakistan. Sci Total Environ 685:659–666
82. Jwaideh MA, Sutanudjaja EH, Dalin C (2022) Global impacts of nitrogen and phosphorus fertiliser use for major crops on aquatic biodiversity. Int J Life Cycle Assess 27(8):1058–1080
83. Ludemann CI, Gruere A, Heffer P, Dobermann A (2022) Global data on fertilizer use by crop and by country. Sci Data 9(1):501
84. Ma L, Zhang Y, Chen S, Yu L, Zhu Y (2022) Environmental effects and their causes of agricultural production: evidence from the farming regions of China. Ecol Ind 144:109549
85. Brackin R, Buckley S, Pirie R, Visser F (2019) Predicting nitrogen mineralisation in Australian irrigated cotton cropping systems. Soil Res 57(3):247–256
86. Harris SJ, Cendón DI, Hankin SI, Peterson MA, Xiao S, Kelly BF (2022) Isotopic evidence for nitrate sources and controls on denitrification in groundwater beneath an irrigated agricultural district. Sci Total Environ 817:152606
87. Yallop O (2014) Citarum, the most polluted river in the world? The Telegraph, 11 Dec 2014. https://www.telegraph.co.uk/news/worldnews/asia/indonesia/11285756/Citarum-the-most-polluted-river-in-the-world.html
88. Intergovernmental Panel on Climate Change (IPCC) (2019) 2019 refinement to the 2006 IPCC guidelines for national greenhouse gas inventories: agriculture, forestry and other land use, vol 4. IPCC, p 824
89. Zhang WF, Dou ZX, He P, Ju XT, Powlson D, Chadwick D, Zhang FS (2013) New technologies reduce greenhouse gas emissions from nitrogenous fertilizer in China. Proc Natl Acad Sci 110(21):8375–8380
90. Brentrup F, Hoxha A, Christensen B (2016) Carbon footprint analysis of mineral fertilizer production in Europe and other world regions. In: Proceedings of the 10th international conference on life cycle assessment of food (LCA food 2016), vol 19

91. Walling E, Vaneeckhaute C (2020) Greenhouse gas emissions from inorganic and organic fertilizer production and use: a review of emission factors and their variability. J Environ Manage 276:111211
92. Cui Z, Zhang H, Chen X, Zhang C, Ma W, Huang C, Dou Z (2018) Pursuing sustainable productivity with millions of smallholder farmers. Nature 555(7696):363–366
93. Liu X, Vitousek P, Chang Y, Zhang W, Matson P, Zhang F (2016) Evidence for a historic change occurring in China. Environ Sci Technol 50(6):505–506
94. Xia L, Lam SK, Chen D, Wang J, Tang Q, Yan X (2017) Can knowledge-based N management produce more staple grain with lower greenhouse gas emission and reactive nitrogen pollution? A meta-analysis. Glob Change Biol 23(5):1917–1925
95. Tao F, Palosuo T, Valkama E, Mäkipää R (2019) Cropland soils in China have a large potential for carbon sequestration based on literature survey. Soil Till Res 186:70–78
96. Sandin G, Roos S, Spak B, Zamani B, Peters G (2019) Environmental assessment of Swedish clothing consumption—six garments, sustainable futures. Mistra Future Fashion, 6
97. van der Velden NM, Patel MK, Vogtländer JG (2014) LCA benchmarking study on textiles made of cotton, polyester, nylon, acryl, or elastane. Int J Life Cycle Assess 19:331–356
98. Zhang Y, Liu X, Xiao R, Yuan Z (2015) Life cycle assessment of cotton T-shirts in China. Int J Life Cycle Assess 20:994–1004
99. Tun T, Kua HW (2015) Integrated policy design through life cycle sustainability assessment: a case study of cotton garments. In: Sustainability matters: environmental and climate changes in the Asia-Pacific. Springer, pp 303–330
100. Schmutz M, Hischier R, Som C (2021) Factors allowing users to influence the environmental performance of their T-shirt. Sustainability 13(5):2498
101. Moazzem S, Daver F, Crossin E, Wang L (2018) Assessing environmental impact of textile supply chain using life cycle assessment methodology. J Text Inst 109(12):1574–1585
102. Fidan FS, Aydogan EK, Uzal N (2022) The impact of organic cotton use and consumer habits in the sustainability of jean production using the LCA approach. Environ Sci Pollut Res 30(4):8853–8867
103. International Energy Agency (2016) Energy, climate change and environment: 2016 insights. International Energy Agency, p 113
104. Bartl A (2020) Textiles production and end-of-life management options. In: Plastic waste and recycling. Academic Press, pp 251–279
105. Eunomia (2015) The potential contribution of waste management to a low carbon economy. https://eunomia.eco/reports/the-potential-contribution-of-waste-management-to-a-low-carbon-economy/
106. Pakula C, Stamminger R (2010) Electricity and water consumption for laundry washing by washing machine worldwide. Energy Effic 3:365–382
107. DuPont (2013) Consumer laundry study. https://www.scribd.com/document/248386105/DuPont-Indian-Laundry-Study-Infographic-pdf
108. World Bank, AQUASTAT, & FAO (2014) Dataset: annual freshwater withdrawals, total. World Bank. https://data.worldbank.org/indicator/ER.H2O.FWIN.ZS
109. Reig P, Gassert F, Luo T, Maddocks A (2013) Water stress by country. Water Resources Institute. https://www.wri.org/data/water-stress-country
110. Klint E, Peters G (2021) Sharing is caring-the importance of capital goods when assessing environmental impacts from private and shared laundry systems in Sweden. Int J Life Cycle Assess 26:1085–1099
111. Portmann FT, Siebert S, Döll P (2010) MIRCA2000—global monthly irrigated and rainfed crop areas around the year 2000: a new high-resolution data set for agricultural and hydrological modeling. Glob Biogeochem Cycles 24(1)
112. Mekonnen MM, Hoekstra AY (2020) Sustainability of the blue water footprint of crops. Adv Water Resour 143:103679
113. Mekonnen MM, Hoekstra AY (2011) The green, blue and grey water footprint of crops and derived crop products. Hydrol Earth Syst Sci 15(5):1577–1600. https://doi.org/10.5194/hess-15-1577-2011

114. Gustafson DI, Collins M, Fry J, Smith S, Matlock M, Zilberman D, Ramsey N (2014) Climate adaptation imperatives: global sustainability trends and eco-efficiency metrics in four major crops—canola, cotton, maize, and soybeans. Int J Agric Sustain 12(2):146–163
115. Liu S, Luo G, Wang H (2020) Temporal and spatial changes in crop water use efficiency in Central Asia from 1960 to 2016. Sustainability 12(2):572
116. Blanc E, Caron J, Fant C, Monier E (2017) Is current irrigation sustainable in the United States? An integrated assessment of climate change impact on water resources and irrigated crop yields. Earth's Future 5(8):877–892
117. Jans Y, von Bloh W, Schaphoff S, Müller C (2021) Global cotton production under climate change—implications for yield and water consumption. Hydrol Earth Syst Sci 25(4):2027–2044
118. Bonetti S, Sutanudjaja EH, Mabhaudhi T, Slotow R, Dalin C (2022) Climate change impacts on water sustainability of South African crop production. Environ Res Lett 17(8):084017
119. Li X, Ren J, Wu Z, Wu X, Ding X (2021) Development of a novel process-level water footprint assessment for textile production based on modularity. J Clean Prod 291:125884
120. Sherrington C, Darrah C, Hann S, Cole G, Corbin M (2016) Study to support the development of measures to combat a range of marine litter sources. Report No. 410. European Commission DG Environment
121. ISO (2006) 14040. Environmental management—life cycle assessment—principles and framework. International Organization for Standardization, pp 235–248
122. Kazan H, Akgul D, Kerc A (2020) Life cycle assessment of cotton woven shirts and alternative manufacturing techniques. Clean Technol Environ Policy 22:849–864
123. Morita AM, Moore CCS, Nogueira AR, Kulay L, Ravagnani MADSS (2020) Assessment of potential alternatives for improving environmental trouser jeans manufacturing performance in Brazil. J Clean Prod 247:119156
124. Munasinghe P, Druckman A, Dissanayake DGK (2021) A systematic review of the life cycle inventory of clothing. J Clean Prod 320:128852
125. Munasinghe P (2024) The use of ISO 14062:2002 sustainable design strategies of the designers to address SDG 12: ensure sustainable consumption and production patterns in the export apparel industry in Sri Lanka. In: Ranathunga GM (ed) Textiles of Sri Lanka. SDGs and textiles. Springer, Singapore. https://doi.org/10.1007/978-981-97-0569-6_7
126. Yu Z, Yang Y (2025) Carbon footprint of global cotton production. Resour Environ Sustain 20:100214
127. Yalcin-Enis I, Kucukali-Ozturk M, Sezgin H (2019) Risks and management of textile waste. In: Nanoscience and biotechnology for environmental applications. Springer, pp 29–53
128. Umarhojiyev H (2025) Circular economy in textile manufacturing: AI-driven approaches to waste reduction. Am J Educ Learn 3(4):454–460
129. Bansal R (2023) Design disruption: crafting a technology that eliminates fabric waste. Texfash. https://texfash.com/special/design-disruption-crafting-a-technology-that-eliminates-fabric-waste
130. Coats Digital (2022) Case study: Duc Hanh Garment JSC—IntelloCut. https://www.coatsdigital.com/wp-content/uploads/2022/01/Coats_Digital_Case_Study_IntelloCut_Duc_Hanh_English.pdf
131. Forde M (2019) Company of the year: Nike. Supply Chain Dive. https://www.supplychaindive.com/news/nike-supply-chain-Celect-dive-awards/566234/
132. Roland Berger (2024) From waste to worth: fashion industry report. https://www.rolandberger.com/publications/publication_pdf/Roland-Berger_From-waste-to-worth_Fashion.pdf
133. Michel S (2025) AI innovation driving water efficiency in textile manufacturing. Textile Network. https://textile-network.com/en/Fashion/Fertigung/AI-innovation-driving-water-efficiency-in-textile-manufacturing
134. AltEnergyMag (2024) Implementing AI for energy efficiency in manufacturing. https://www.altenergymag.com/story/2024/09/implementing-ai-for-energy-efficiency-in-manufacturing/43312/

135. NIST (2022) Improving resource efficiency for the textile industry. Workshop report. https://www.nist.gov/mml/mmsd/security-technologies-group/improving-resource-efficiency-textile-industry
136. HM Foundation & HKRITA (2024) AI-powered smart garment sorting system. https://hmfoundation.com/project/ai-powered-smart-garment-sorting-system-automating-textile-recycling/
137. Refiberd (2025) Intelligent sorting for textile-to-textile recycling. https://refiberd.com/
138. Du W, Zheng J, Li W, Liu Z, Wang H, Han X (2022) Efficient recognition and automatic sorting technology of waste textiles based on online near infrared spectroscopy and convolutional neural network. Resour Conserv Recycl 180:106157
139. Cura K, Rintala N, Kamppuri T, Saarimäki E, Heikkilä P (2021) Textile recognition and sorting for recycling at an automated line using near infrared spectroscopy. Recycling 6(1):11
140. RecyclingInside (2023) Recycling post-industrial waste in the textile industry: a circular future. https://recyclinginside.com/recycling-technology/separation-and-sorting-technology/recycling-post-industrial-waste-in-the-textile-industry-a-circular-future/
141. Wastex (2023) Textile sorting technologies. https://www.wastexrecycling.com/
142. Wu Y, Xu Z, Yang W, Ning Z, Dong H (2022) Review on the application of hyperspectral imaging technology of the exposed cortex in cerebral surgery. Front Bioeng Biotechnol 10:906728
143. Holt GA, Simonton JL, Beruvides MG, Canto A (2004) Utilization of cotton gin by-products for the manufacturing of fuel pellets: an economic perspective. Appl Eng Agric 20(4):423–430
144. Agblevor FA, Cundiff JS, Mingle C, Li W (2006) Storage and characterization of cotton gin waste for ethanol production. Resour Conserv Recycl 46(2):198–216
145. Silverstein RA, Chen Y, Sharma-Shivappa RR, Boyette MD, Osborne J (2007) A comparison of chemical pretreatment methods for improving saccharification of cotton stalks. Bioresour Technol 98(16):3000–3011
146. Jeoh T, Agblevor FA (2001) Characterization and fermentation of steam exploded cotton gin waste. Biomass Bioenergy 21(2):109–120
147. Demirbaş A (1998) Yields of oil products from thermochemical biomass conversion processes. Energy Convers Manage 39(7):685–690
148. Weiming Y, Xueyuan B, Fang H, Zhihe L, Yongjun L, Shuangning X (2005) Bio-oil from agricultural residues by fast pyrolysis. In: 2005 ASAE annual meeting. American Society of Agricultural and Biological Engineers, p 1
149. Li LIN, Zhang H (2004) Preparing levoglucosan derived from waste material by pyrolysis. Energy Sources 26(11):1053–1059
150. Chen G, Leung DYC (2003) Experimental investigation of biomass waste, (rice straw, cotton stalk, and pine sawdust), pyrolysis characteristics. Energy Sources 25(4):331–337
151. Demirbaş A (2001) Yields of hydrogen-rich gaseous products via pyrolysis from selected biomass samples. Fuel 80(13):1885–1891
152. Demirbas A, Arin G (2004) Hydrogen from biomass via pyrolysis: relationships between yield of hydrogen and temperature. Energy Sources 26(11):1061–1069
153. Hansen KH, Angelidaki I, Ahring BK (1998) Anaerobic digestion of swine manure: inhibition by ammonia. Water Res 32(1):5–12
154. Funk PA, Armijo CB, Hanson AT, Samani ZA, Macias-Corral MA, Smith GB, Riordan JT (2005) Converting gin and dairy wastes to methane. Trans ASAE 48(3):1197–1201
155. Macias-Corral MA, Samani ZA, Hanson AT, DelaVega R, Funk PA (2005) Producing energy and soil amendment from dairy manure and cotton gin waste. Trans ASAE 48(4):1521–1526
156. Kelleher BP, Leahy JJ, Henihan AM, O'dwyer TF, Sutton D, Leahy MJ (2002) Advances in poultry litter disposal technology—a review. Bioresour Technol 83(1):27–36
157. Yousef S, Tatariants M, Tichonovas M, Sarwar Z, Jonuškienė I, Kliucininkas L (2019) A new strategy for using textile waste as a sustainable source of recovered cotton. Resour Conserv Recycl 145:359–369
158. Ütebay B, Çelik P, Çay A (2019) Effects of cotton textile waste properties on recycled fibre quality. J Clean Prod 222:29–35

159. U.S. Environmental Protection Agency (2019) Facts and figures about materials, waste and recycling
160. Smits H (2018) The time is now for recycled cotton—and here's why. Sourcing J 5:89–96
161. Miljögiraff (2016) LCA of recycling cotton. Report 75 [Miljögiraff report]. Miljögiraff, Gothenburg, Sweden
162. European Commission (2020) EU citizenship report 2020: empowering citizens and protecting their rights, 15 Dec 2020. Directorate-General for Justice and Consumers. https://op.europa.eu/en/publication-detail/-/publication/b89eb334-3ed0-11eb-b27b-01aa75ed71a1/language-en
163. Renewcell AB (2019) Renewcell—we make fashion circular. https://renewcell.com/
164. European Commission (2019) TeKiDe technology creates sustainable new textiles from waste cotton in Helsinki-Uusimaa. https://ec.europa.eu/regional_policy/en/projects/finland/tekide-technology-creates-sustainable-new-textiles-from-waste-cotton-in-helsinki-uusimaa
165. Esteve-Turrillas FA, de La Guardia M (2017) Environmental impact of recover cotton in textile industry. Resour Conserv Recycl 116:107–115
166. H&M Group (2019) Sustainability report 2018. Stockholm, Sweden
167. Maciel MMÁD, de Carvalho Benini KCC, Voorwald HJC, Cioffi MOH (2019) Obtainment and characterization of nanocellulose from an unwoven industrial textile cotton waste: effect of acid hydrolysis conditions. Int J Biol Macromol 126:496–506
168. De Silva R, Byrne N (2017) Utilization of cotton waste for regenerated cellulose fibres: influence of degree of polymerization on mechanical properties. Carbohydr Polym 174:89–94
169. Hou W, Ling C, Shi S, Yan Z, Zhang M, Zhang B, Dai J (2018) Separation and characterization of waste cotton/polyester blend fabric with hydrothermal method. Fibers Polym 19:742–750
170. Björquist S, Aronsson J, Henriksson G, Persson A (2018) Textile qualities of regenerated cellulose fibers from cotton waste pulp. Text Res J 88(21):2485–2492
171. Muthu SS (ed) (2014b) End-of-life management of textile products. Woodhead Publishing, pp 144–162
172. Leal Filho W, Ellams D, Han S, Tyler D, Boiten VJ, Paço A, Balogun AL (2019) A review of the socio-economic advantages of textile recycling. J Clean Prod 218:10–20
173. Christoph R, Schmidt B, Steinberner U, Dilla W, Karinen R (2000) Glycerol. In: Ullmann's encyclopedia of industrial chemistry. Wiley VCH
174. Sandin G, Peters GM (2018) Environmental impact of textile reuse and recycling—a review. J Clean Prod 184:353–365
175. Ferrão P, Amaral J (2006) Design for recycling in the automobile industry: new approaches and new tools. J Eng Des 17(5):447–462
176. Perry N, Bernard A, Laroche F, Pompidou S (2012) Improving design for recycling—application to composites. CIRP Ann 61(1):151–154
177. Payne A (2015) Open-and closed-loop recycling of textile and apparel products. In: Handbook of life cycle assessment (LCA) of textiles and clothing. Woodhead Publishing, pp 103–123
178. Korhonen MR, Dahlbo H (2007) Reducing greenhouse gas emissions by recycling plastics and textiles into products. Finnish Environment Institute, Helsinki, Finland
179. Satyanarayana KG, Arizaga GG, Wypych F (2009) Biodegradable composites based on lignocellulosic fibers—an overview. Prog Polym Sci 34(9):982–1021
180. Rajput D, Bhagade SS, Raut SP, Ralegaonkar RV, Mandavgane SA (2012) Reuse of cotton and recycle paper mill waste as building material. Constr Build Mater 34:470–475
181. Ardanuy M, Claramunt J, Toledo Filho RD (2015) Cellulosic fiber reinforced cement-based composites: a review of recent research. Constr Build Mater 79:115–128
182. Nikolić S, Lazić V, Veljović Đ, Mojović L (2017) Production of bioethanol from pre-treated cotton fabrics and waste cotton materials. Carbohydr Polym 164:136–144
183. Hong F, Guo X, Zhang S, Han SF, Yang G, Jönsson LJ (2012) Bacterial cellulose production from cotton-based waste textiles: enzymatic saccharification enhanced by ionic liquid pretreatment. Bioresour Technol 104:503–508
184. Habibi Y, Lucia LA, Rojas OJ (2010) Cellulose nanocrystals: chemistry, self-assembly, and applications. Chem Rev 110(6):3479–3500

185. Ghasemian M, Zilouei H, Asadinezhad A (2016) Enhanced biogas and biohydrogen production from cotton plant wastes using alkaline pretreatment. Energy Fuels 30(12):10484–10493
186. Ranjithkumar M, Ravikumar R, Sankar MK, Kumar MN, Thanabal V (2017) An effective conversion of cotton waste biomass to ethanol: a critical review on pretreatment processes. Waste Biomass Valorization 8:57–68
187. Jeihanipour A, Taherzadeh MJ (2009) Ethanol production from cotton-based waste textiles. Bioresour Technol 100(2):1007–1010
188. Jeihanipour A, Karimi K, Niklasson C, Taherzadeh MJ (2010) A novel process for ethanol or biogas production from cellulose in blended-fibers waste textiles. Waste Manage 30(12):2504–2509
189. Sharma-Shivappa RR, Chen Y (2008) Conversion of cotton wastes to bioenergy and value-added products. Trans ASABE 51(6):2239–2246
190. Gwenzi W, Chaukura N, Noubactep C, Mukome FN (2017) Biochar-based water treatment systems as a potential low-cost and sustainable technology for clean water provision. J Environ Manage 197:732–749
191. Barrow CJ (2012) Biochar: potential for countering land degradation and for improving agriculture. Appl Geogr 34:21–28
192. Rago YP, Surroop D, Mohee R (2018) Assessing the potential of biofuel (biochar) production from food wastes through thermal treatment. Bioresour Technol 248:258–264
193. Zhou X, Wang P, Zhang Y, Zhang X, Jiang Y (2016) From waste cotton linter: a renewable environment-friendly biomass based carbon fibers preparation. ACS Sustain Chem Eng 4(10):5585–5593
194. Wang B, Karthikeyan R, Lu XY, Xuan J, Leung MK (2013) Hollow carbon fibers derived from natural cotton as effective sorbents for oil spill cleanup. Ind Eng Chem Res 52(51):18251–18261
195. Zheng J, Zhao Q, Ye Z (2014) Preparation and characterization of activated carbon fiber (ACF) from cotton woven waste. Appl Surf Sci 299:86–91
196. Ekrami E, Dadashian F, Arami M (2016) Adsorption of methylene blue by waste cotton activated carbon: equilibrium, kinetics, and thermodynamic studies. Desalin Water Treat 57(15):7098–7108
197. Wanassi B, Hariz IB, Ghimbeu CM, Vaulot C, Hassen MB, Jeguirim M (2017) Carbonaceous adsorbents derived from textile cotton waste for the removal of Alizarin S dye from aqueous effluent: kinetic and equilibrium studies. Environ Sci Pollut Res 24:10041–10055
198. Arena N, Lee J, Clift R (2016) Life cycle assessment of activated carbon production from coconut shells. J Clean Prod 125:68–77
199. Nunes LJ, Godina R, Matias JC, Catalão JP (2018) Economic and environmental benefits of using textile waste for the production of thermal energy. J Clean Prod 171:1353–1360
200. Breault RW (2010) Gasification processes old and new: a basic review of the major technologies. Energies 3(2):216–240
201. Pathak S (2019) Biomimicry: (innovation inspired by nature). Int J New Technol Res 5:1–4
202. Ütebay B, Çelik P, Çay A (2020) Waste in textile and leather sectors. In: Textile wastes: status and perspectives. IntechOpen, pp 1–20
203. Ribul M, Lanot A, Pisapia CT, Purnell P, McQueen-Mason SJ, Baurley S (2021) Mechanical, chemical, biological: moving towards closed-loop bio-based recycling in a circular economy of sustainable textiles. J Clean Prod 326:129325
204. Xu B, Chen Q, Fu B, Zheng R, Fan J (2022) Current situation and construction of recycling system in China for post-consumer textile waste. Sustainability 14(24):16635
205. Schumacher KA, Forster AL (2022) Textiles in a circular economy: an assessment of the current landscape, challenges, and opportunities in the United States. Front Sustain 3:1038323
206. Fan W, Wang Y, Liu R, Zou J, Yu X, Liu Y, Meng J (2024) Textile production by additive manufacturing and textile waste recycling: a review. Environ Chem Lett 22(4):1929–1987
207. Chopra SS, Dong L, Kaur G, Len C, Lin CSK (2023) Sustainable process design for circular fashion: advances in sustainable chemistry for textile waste valorisation. Curr Opin Green Sustain Chem 39:100747

208. Akter MMK, Haq UN, Islam MM, Uddin MA (2022) Textile-apparel manufacturing and material waste management in the circular economy: a conceptual model to achieve sustainable development goal (SDG) 12 for Bangladesh. Clean Environ Syst 4:100070
209. Shirvanimoghaddam K, Motamed B, Ramakrishna S, Naebe M (2020) Death by waste: fashion and textile circular economy case. Sci Total Environ 718:137317
210. Tang KHD (2023) State of the art in textile waste management: a review. Textiles 3(4):454–467
211. Tisserant A, Pauliuk S, Merciai S, Schmidt J, Fry J, Wood R, Tukker A (2017) Solid waste and the circular economy: a global analysis of waste treatment and waste footprints. J Ind Ecol 21(3):628–640
212. Juanga-Labayen JP, Labayen IV, Yuan Q (2022) A review on textile recycling practices and challenges. Textiles 2(1):174–188
213. Yaseen DA, Scholz M (2018) Treatment of synthetic textile wastewater containing dye mixtures with microcosms. Environ Sci Pollut Res 25:1980–1997
214. Fortuna LM, Diyamandoglu V (2017) Optimization of greenhouse gas emissions in second-hand consumer product recovery through reuse platforms. Waste Manage 66:178–189
215. Islam S (2021) Waste management strategies in fashion and textiles industry: challenges are in governance, materials culture and design-centric. In: Waste management in the fashion and textile industries. Woodhead Publishing, pp 275–293
216. Cesaro A, Conte A, Belgiorno V, Siciliano A, Guida M (2019) The evolution of compost stability and maturity during the full-scale treatment of the organic fraction of municipal solid waste. J Environ Manage 232:264–270
217. Moorhouse D, Moorhouse D (2017) Sustainable design: circular economy in fashion and textiles. Des J 20(sup1):S1948–S1959
218. Palme A, Peterson A, de la Motte H, Theliander H, Brelid H (2017) Development of an efficient route for combined recycling of PET and cotton from mixed fabrics. Text Cloth Sustain 3:1–9
219. Zach J, Hroudová J, Korjenic A (2016) Environmentally efficient thermal and acoustic insulation based on natural and waste fibers. J Chem Technol Biotechnol 91(8):2156–2161
220. Massoudinejad M, Amanidaz N, Santos RM, Bakhshoodeh R (2019) Use of municipal, agricultural, industrial, construction and demolition waste in thermal and sound building insulation materials: a review article. J Environ Health Sci Eng 17:1227–1242
221. Baloyi RB, Gbadeyan OJ, Sithole B, Chunilall V (2024) Recent advances in recycling technologies for waste textile fabrics: a review. Text Res J 94(3–4):508–529
222. Kadirvelu K, Thamaraiselvi K, Namasivayam C (2001) Removal of heavy metals from industrial wastewaters by adsorption onto activated carbon prepared from an agricultural solid waste. Bioresour Technol 76(1):63–65
223. Amaya A, Medero N, Tancredi N, Silva H, Deiana C (2007) Activated carbon briquettes from biomass materials. Bioresour Technol 98(8):1635–1641
224. Attia AA, El-Hendawy ANA, Khedr SA, El-Nabarawy TH (2004) Textural properties and adsorption of dyes onto carbons derived from cotton stalks. Adsorp Sci Technol 22(5):411–426
225. Thangamani KS, Sathishkumar M, Sameena Y, Vennilamani N, Kadirvelu K, Pattabhi S, Yun SE (2007) Utilization of modified silk cotton hull waste as an adsorbent for the removal of textile dye (reactive blue MR) from aqueous solution. Bioresour Technol 98(6):1265–1269
226. Kadirvelu K, Kavipriya M, Karthika C, Radhika M, Vennilamani N, Pattabhi S (2003) Utilization of various agricultural wastes for activated carbon preparation and application for the removal of dyes and metal ions from aqueous solutions. Bioresour Technol 87(1):129–132
227. Hashem A, Aly AA, Aly AS, Hebeish A (2006) Quaternization of cotton stalks and palm tree particles for removal of acid dye from aqueous solutions. Polym-Plast Technol Eng 45(3):389–394
228. McKay G, Porter JF, Prasad GR (1999) The removal of dye colours from aqueous solutions by adsorption on low-cost materials. Water Air Soil Pollut 114:423–438
229. Rossi J, Stewart L (2006) Using cotton byproducts in beef cattle diets. Bulletin 1311
230. Pordesimo LO, Ray SJ, Buschermohle MJ, Waller JC, Wilkerson JB (2005) Processing cotton gin trash to enhance in vitro dry matter digestibility in reduced time. Bioresour Technol 96(1):47–53

231. Philippoussis A, Zervakis G, Diamantopoulou P (2001) Bioconversion of agricultural lignocellulosic wastes through the cultivation of the edible mushrooms *Agrocybe aegerita*, *Volvariella volvacea* and *Pleurotus* spp. World J Microbiol Biotechnol 17:191–200
232. Zhang YHP, Himmel ME, Mielenz JR (2006) Outlook for cellulase improvement: screening and selection strategies. Biotechnol Adv 24(5):452–481
233. Seyis I, Aksoz N (2005) Xylanase production from *Trichoderma harzianum* 1073 D3 with alternative carbon and nitrogen sources. Food Technol Biotechnol 43(1):37–40
234. Paredes C, Bernal MP, Roig A, Cegarra J (2001) Effects of olive mill wastewater addition in composting of agroindustrial and urban wastes. Biodegradation 12(4):225–234
235. Paredes C, Bernal MP, Cegarra J, Roig A (2002) Bio-degradation of olive mill wastewater sludge by its co-composting with agricultural wastes. Bioresour Technol 85(1):1–8
236. García-Gómez A, Bernal MP, Roig A (2005) Organic matter fractions involved in degradation and humification processes during composting. Compost Sci Util 13(2):127–135
237. Tejada M, Gonzalez JL (2006) Crushed cotton gin compost on soil biological properties and rice yield. Eur J Agron 25(1):22–29
238. Holt G, Buser M, Harmel D, Potter K, Pelletier M (2005a) Comparison of cotton-based hydro-mulches and conventional wood and paper hydro-mulches—study 1. J Cotton Sci 9(3):121–127
239. Holt G, Buser M, Harmel D, Potter K (2005b) Comparison of cotton-based hydro-mulches and conventional wood and paper hydro-mulches—study 2. J Cotton Sci 9(3):128–134
240. Guler C, Ozen R (2004) Some properties of particleboards made from cotton stalks (*Gossypium hirsitum* L.). Holz als Roh- und Werkstoff 62:40–43

Selection, Measurement, Evaluation and Reporting Method of Mitigation Actions Applied to Sustainable Urban Freight Transport Sector

Tássia Faria de Assis, Victor Hugo Souza de Abreu, and Marcio de Almeida D'Agosto

Abstract In alignment with global efforts to achieve the climate targets set by the Paris Agreement and mitigate environmental impacts, the urban freight transport (UFT) sector must undergo significant decarbonization to help reduce carbon footprint and consequently curb global temperature rise. Reducing greenhouse gas (GHG) emissions in UFT demands a multifaceted approach, including optimizing logistics and operational efficiency, transitioning to low-carbon and zero-emission transport modes, and adopting advanced fuel-efficient technologies. This chapter presents the Selection, Measurement, Evaluation and Reporting (SMER) method, a important tool for evaluating and tracking the effectiveness of these strategies and developed in the study of Assis (Desenvolvimento de um Método para Selecionar, Medir, Avaliar e Reportar Boas Práticas para promover a Logística Sustentável. Doctoral thesis submitted to the Graduate Program in Transportation Engineering, COPPE, Federal University of Rio de Janeiro, 2022 [11]). The SMER framework enables companies and policymakers to systematically measure and select sustainability initiatives, evaluate progress, transparently report results, and ensure the verification of emission reductions by enhancing the MRV method. Furthermore, it plays a critical role in supporting the implementation and assessment of Nationally Determined Contributions (NDCs) by providing a structured approach to evaluating three key dimensions: GHG emissions quantification, effectiveness of mitigation actions, and tracking of financial, technological, and capacity-building support. By establishing a robust SMER method, the UFT sector can enhance accountability, drive continuous improvement, and accelerate the transition toward a low-carbon economy.

T. F. de Assis (✉) · M. de Almeida D'Agosto
Transport Engineering Programme (Programa de Engenharia de Transportes—PET), COPPE-UFRJ, Federal University of Rio de Janeiro (Universidade Federal do Rio de Janeiro—UFRJ), Rio de Janeiro, Brazil
e-mail: tassiafa@pet.coppe.ufrj.br

V. H. S. de Abreu
Transport Engineering Department (Departamento de Engenharia de Transportes—DET), POLI-UFRJ, Federal University of Rio de Janeiro (Universidade Federal do Rio de Janeiro—UFRJ), Rio de Janeiro, Brazil

S. S. Muthu (ed.), *Energy Efficiency and Carbon Footprint Assessment*, Environmental Footprints and Eco-design of Products and Processes,
https://doi.org/10.1007/978-3-032-10909-5_3

Keywords Urban freight transport · Best practice · Sustainable method · Sustainability · Carbon footprint · MRV method

1 Introduction

Enterprises that need to distribute their products to numerous customers and businesses in urban areas globally aim to remain competitive and sustainable [10, 28]. To accomplish this, they seek to enhance the efficiency of urban freight transport (UFT) by lowering costs, improving service levels, and reducing environmental impacts through policies that optimize UFT operations [18, 39]. Additionally, for each country to meet its Nationally Determined Contributions (NDCs) established by the Paris Agreement in 2015, the freight transport sector must undergo a substantial process of decarbonization by 2050. This transition is essential to limit the global temperature rise to below 2 °C. Achieving this requires the implementation of robust and effective mitigation policies at the governmental level.

According to the Intergovernmental Panel on Climate Change [50], transport-related emissions reached around 8.7 gigatons of CO_2-eq in 2019, accounting for 23% of global energy-related CO_2 emissions. Road transport represented nearly 70% of these emissions. Without stronger and sustained mitigation policies, projections indicate that emissions from this sector could continue to grow and may approach 12 gigatons of CO_2 by 2050.

Direct greenhouse gas (GHG) emissions from the transport sector can be mitigated through strategic business policies aimed at improving operational efficiency. These strategies include optimizing activity levels, shifting to lower-carbon transport modes, reducing energy intensity, and decreasing the carbon intensity of fuels. In this context, methods to section, measurement, evaluation and reporting play a crucial role in guiding the selection and evaluation of mitigation actions.

Many enterprises have explored different approaches to improving their business models, implementing individual solutions to enhance operational sustainability. By adopting Best Practices, these companies can scale up and further develop their sustainability efforts [22]. To facilitate the integration of best practices into the UFT sector, this chapter proposes a structured procedure designed to assist enterprises in incorporating sustainable strategies into their operations. The proposed method enables companies to select, measure, monitor, report, and verify their mitigation efforts through the Selection, Measurement, Evaluation and Reporting (SMER) framework.

2 Sustainable Methods to Selection, Measurement, Evaluation and Reporting

Urban logistics is vital for economic development, but it causes serious environmental impacts such as GHG emissions, congestion and noise pollution [4]. Adopting sustainable strategies is crucial to mitigating these effects, promoting energy efficiency and a better quality of urban life [7].

However, there are still methodological challenges in selecting, measuring, monitoring and reporting on these strategies, which makes it necessary to implement strong methods for evaluating sustainable strategies in UFT to ensure that the initiatives adopted are effective and measurable. Studies such as Anand et al. [4] and Jamshidi et al. [51] highlight that the standardization of metrics, such as CO_2 emissions, fuel consumption and the energy efficiency of routes, makes it possible to compare different approaches and identify the most promising ones. In addition, continuous monitoring helps to adjust policies and technologies in real-time, while transparent reporting makes it easier to be accountable to stakeholders and regulators [69].

Despite progress, methodological challenges persist and one of the main gaps is the lack of standardization in sustainability indicators, which makes it difficult to compare studies and initiatives [41, 48]. Furthermore, many methods do not adequately consider contextual factors, such as regional differences in infrastructure and logistics demand. Another problem is the scarcity of frameworks that incorporate social aspects of sustainability, such as equity in the distribution of logistics benefits [38]. To provide an overview of examples of current methods applied to sustainability issues, this section discusses methods identified in the literature.

2.1 Selection Methods

The selection of good sustainable practices in UFT can be guided by maturity models, which help companies to assess and progressively improve their practices. These models provide a sequential structure, divided into levels, which allows the evolution towards sustainable logistics to be measured, considering factors such as uncertainties, complexity and market demands [23].

One example is the Sustainable Logistics Maturity Model (SLMM), proposed by Werner-Lewandowska and Golinska-Dawson [73], which assesses companies' ability to adopt logistics strategies aligned with the economic, environmental and social dimensions. This model makes it possible to measure the performance of service providers, identifying gaps and promoting the implementation of sustainable tools and practices.

The SLMM model, alone or combined with other maturity models such as the Technology Readiness Level (TRL), to align technological adoption with economic, environmental and social sustainability goals [73]. TRL, developed by NASA, can evaluate sustainable urban transport strategies, in which it rates technologies from 1 to 9, with low levels being initial concepts and high levels being solutions ready for implementation at scale [54, 56].

TRL makes it possible to identify the stage of development of sustainable technologies (e.g. electric vehicles, intelligent routing systems, alternative fuels), helping managers to prioritize solutions with greater technical and commercial viability [17].

In addition to the methods/models mentioned above, SWOT analysis can be combined with models such as the Sustainable Logistics Maturity Model (SLMM) and Technology Readiness Level (TRL) to align strategies with the company's level of operational and technological maturity [73].

In addition, SWOT (Strengths, Weaknesses, Opportunities, Threats) analysis is also a strategic tool that can be applied to evaluate and select sustainable practices in urban freight transport, allowing managers to identify internal and external factors that impact the effectiveness of green initiatives [45].

The SWOT analysis tool makes it possible to analyze: (i) strengths, identifying competitive advantages such as energy efficiency or sustainable partnerships [14], e.g. electric vehicles or renewable fuels that reduce emissions, (ii) weaknesses, recognizing barriers to implementation such as lack of technical capacity or resistance to change [55], for example, limited infrastructure for recharging or high initial costs, (iii) opportunities, taking advantage of trends such as favorable regulations [30], for example, tax incentives for low-carbon vehicles or growing demand for green logistics, and (iv) threats, anticipating risks such as changes in public policy or unsustainable competition [61], for example, volatility in the price of alternative fuels or urban congestion.

2.2 *Measurement Method*

To assess and improve sustainable performance in this area, it is essential to adopt measurement methods, both in ex-ante scenarios, i.e. those that take place before strategies are implemented, and ex-post scenarios, which occur after implementation [31, 65].

Companies can use methods that assess potential impacts and set targets based on scientific data, such as the Science-Based Targets Initiative (SBTi). This initiative encourages the setting of carbon reduction targets to limit global warming to less than 2 °C above pre-industrial levels, such targets are developed with common resources and methods, independently assessed and approved by a technical advisory group [37].

Gustavsson [42] analyzed why companies have adopted SBTis, why some include targets for the transport of purchased goods while others do not, how they plan to meet these targets and how the work related to SBTis has influenced the management of GHG emissions, especially in the transport of purchased goods.

It is understood that once sustainable strategies have been implemented, it is essential to measure results and adjust actions since, according to Ambrosio [3], daily logistic operation needs must be evaluated from the type of operations done, load, environment, motivation, monitoring and control of logistic processes. From the adoption of methods such as Measurement, Evaluation and Reporting (MRV). The MRV method is crucial for identifying significant actions from GHG policies [70]. Organizations can only manage what they measure, so MRV tracks progress toward established policy goals.

The importance of measurement, reporting, and verification (MRV) increased after COP 16 in 2010, which emphasized result-based compensatory phase implementation [58]. According to WRI [72], MRV is essential for implementing NDCs, which outline countries' mitigation goals. Measurement identifies emission trends, guides reduction efforts, and monitors progress. Reporting and verification ensure transparency, accountability, and effective resource use.

UNFCCC initially defined M as Measuring but it's now often Monitoring. Reporting presents measured data transparently, and verification assesses its reliability through an independent process [32].

Rich et al. [64] note that MRV standards are used by governments, donor agencies, businesses, and NGOs to measure GHG emissions. WRI [72] divides MRV method into three types: (i) GHG emissions—understanding and reporting an entity's emission profile, (ii) MRV mitigation actions—assessing the effects and implementation of GHG reduction measures, (iii) MRV of support—tracking climate support provision and results.

While focusing on GHG emissions and mitigation in the transport sector, it should also consider non-GHG benefits like safety, mobility, and economic gains [32]. GHG emissions can be measured before policy implementation (ex-ante), during implementation, and after (ex-post) to estimate a policy's impact [64].

2.3 Performance Evaluation Method

In order to promote sustainability by balancing the economic, social and environmental effects of operational activities, it is important to adopt methods such as BSC (Balanced Scorecard), SCOR (Supply Chain Operations Reference), DMAIC (Measure, Analyse, Improve and Control) to help adopt good sustainable practices in order to better assess the logistics performance of business operations [1, 51].

The BSC method is a performance measurement approach that uses financial and non-financial measures to evaluate all aspects of an organization's operations in an integrated manner to ensure that they are linked to the achievement of the organizations' overall objectives [2]. The SBSC method is derived from the BSC method by including environmental and social aspects as essential visions of a sustainable business [59], thus, including the concept and dimension of sustainability in the model has the potential to allow management to meet targets in all three dimensions of sustainability and integrate these dimensions into a single management system rather than parallel systems [44].

SCOR is a model developed by the Supply Chain Council (SCC) to evaluate supply chain management, providing a set of standardized definitions for processes, process elements, and metrics [25]. The structure of the method extends from the first supplier to the final customer in the supply chain, through five process stages: (i) plan; (ii) source; (iii) make; (iv) deliver; and (v) return [29].

The DMAIC, on the other hand, is a method used to Measure, Analyze, Improve, and Control the performance of products, services, and business processes in order to achieve operational excellence [33]. It is applied when the goal is to improve an existing product or service, where, in the analysis phase, the project team uses data analysis tools and process analysis techniques to identify and verify the root causes of the problem [68].

However, although there are methods and tools available to support the adoption of sustainable best practices and to better assess the logistical performance of business operations, it is still important to address the lack of robust methods capable of measuring and analyzing the social and environmental impacts of logistical activities within production systems [16].

2.4 Disclosure and Report Methods

Disclosure of carbon emission results is voluntary, allowing for different interpretations by management. This results in different approaches to measuring and communicating carbon-related information. These approaches in global logistics companies can be attributed to the competing logics of the market and sustainability and can show two different approaches, as theory declarations without strategies actions and showed of practices actions that decreases the carbon emissions in harmony with the society [47]. In studies of Herold and Lee [47], Herold [46], DeWeerdt et al. [27] and Asif et al. [6], research is developed based on cases of companies operating in the logistics and distribution sector in urban areas, using the Carbon Disclosure Project (CDP) method.

The Carbon Disclosure Project (CDP), founded in 2007 in London, United Kingdom, is an independent international non-profit organization that collects and shares data on companies' environmental actions, including the monitoring and reduction of carbon emissions, to promote transparency and sustainability. The method for monitoring companies' environmental actions involves five stages: (1) Disclosure of Greenhouse Gas emissions, (2) Governance through supervision at the board level to determine if there is management-level oversight of climate-related issues in the organization and the appointment of positions or committees responsible for climate-related issues, including incentives for managing these issues, (3) Definition of objectives based on targets set by the Science Based Targets initiative (SBTi) or other targets, (4) Strategic and transition plan, and (5) Target attainment [20].

In the study by Herold and Lee [47], the reports of the companies DHL, FDX and UPS were analyzed to identify the similarities and differences in the information presented in the reports. DeWeerdt et al. [27] also analyzed the carbon management strategies implemented in the DHL company, as well as the measures adopted to reduce its emissions. Asif et al. [6] aimed to provide critical support for sustainability theories and help small and medium-sized enterprises (SMEs) develop strategic business models to meet net-zero emissions objectives and used as an example the realization of emissions-neutral parcel deliveries reported by the MARS group.

In order to apply the CDP method, it is necessary to align guidelines from other standards and evaluate different methods for sustainability reporting aimed at transport. This includes complementary frameworks and standards to support different sets of stakeholders. The main ones are: (i) Global Reporting Initiative (GRI), (ii) Sustainability Accounting Standards Board (SASB), (iii) International Integrated Reporting Council (IIRC), and (iv) Task Force on Climate-Related Financial Disclosures (TCFD) [43, 62].

The Global Reporting Initiative (GRI) was founded in 1977 in Amsterdam, the Netherlands. It is a standard that enables any organization, whether large or small, private or public, to understand and report on its impacts from the economic, environmental and social pillars [36, 53].

The International Integrated Reporting Council (IIRC), founded in 2010 in England, is a global coalition of regulators, investors, companies, standard-setters, accounting professionals, academics and NGOs. It is responsible for recommendations, guidelines and contributions on issues relevant to the organization, including objectives, purpose, vision and mission, as well as strategy and the means to achieve them [21, 71].

The Sustainability Accounting Standards Board (SASB), founded in 2011 in the United States, is a standard responsible for establishing guidelines for the disclosure of material sustainability information by companies, allowing investors to understand how an organization creates business value [66].

The Task Force on Climate-Related Financial Disclosures (TCFD), established in 2017, provides a framework of recommendations for disclosing information pertaining to climate-related issues. Its objective is to enhance decision-making regarding investments, credit, and to improve the understanding of asset concentrations related to carbon emissions [57].

From the methods applied to selection, measure, evaluation and report literature review, the next section proposes a comprehensive procedure to systematically select, measure, monitor, report, and verify best practices. This procedure focuses on identifying strategic and operational mitigation actions that are most appropriate for addressing UFT problems, ensuring effective and measurable solutions.

3 Proposed Methodological Procedure

The procedure aims to guide the select of mitigation actions (best practices) at enterprise strategic level, understood as the elaboration process of visions and missions, specific objectives definition, implementation, evaluation and adoption of corrective

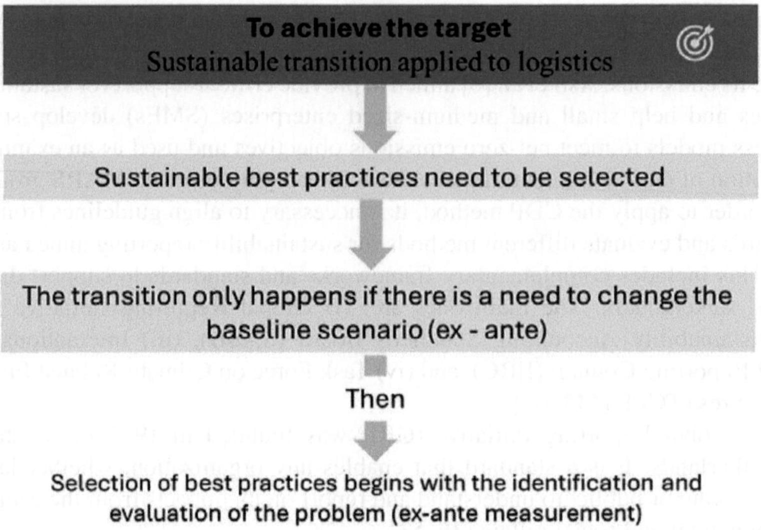

Fig. 1 Logical thinking to build the SMER method. *Source* Own elaboration (2025)

actions [52]. However, the baseline estimate occurs to identify the current scenario of UFT operation in the defined case study, allowing sensitive points analysis to be improved, according to the future expectation. For this method the logical structure is considered presented in Fig. 1.

The procedure steps are as follows: (i) ex-ante measurement (baseline); (ii) selection of best practices for UFT/operation monitoring; (iii) ex-post scenario measurement; (iv) verification of results; and (v) external report. These steps align with the key points shown in Fig. 2, including considerations such as: (1) Which best practices can achieve strategic targets to enhance the sustainability of the UFT sector? (2) How can best practices be effectively measured, monitored, and verified?

3.1 Step 1—Baseline Measurement (Ex-Ante)

At this stage, the baseline is estimated using data derived from historical series and monthly records from the UFT operating system. But it is necessary to delimit the scope of the study, including:

1. Knowing the geographical area

It is necessary to know the location in which the operation takes place, defined by metropolitan region, specific corridors or freight zones, size of the urban area and geographical and environmental characteristics such as relief, climate and temperature, infrastructure quality, etc. [63].

2. Types of vehicles

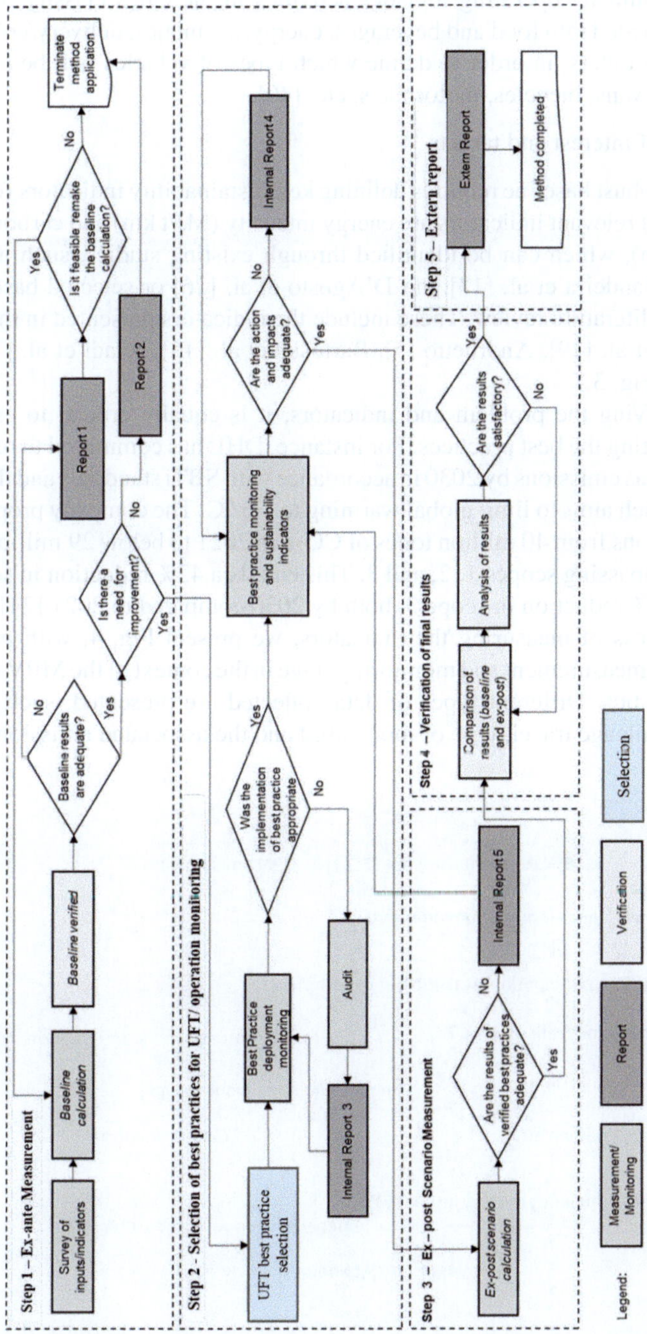

Fig. 2 SMER framework. *Source* Own elaboration (2025) from Assis [11]

It is important to know the type of segment, indicating which types of products or branches the company operating logistics activities in the cities classify itself as, which can be divided into food and beverages, energy, cosmetics, delivery/collection services, among others, in order to define which types of vehicles will be assessed whether trucks, vans, bicycles, motorbikes, etc. [40].

3. Indicators of interest and targets

Establishing a robust baseline requires defining key sustainability indicators for UFT. Among the most relevant indicators are energy intensity (MJ/t km) and carbon intensity (CO_2e/t km), which can be identified through existing studies, such as those conducted by Bandeira et al. [13] and D'Agosto et al. [26] or selected based on a comprehensive literature review. These include the indicators presented in the study of Buldeo Rai et al. [19], Andruetto [5], Bartuška et al. [15], Ayadi et al. [12] and exemplified in Fig. 3.

After identifying the problem and indicators, it is equally crucial to establish targets for selecting the best practices. For instance, DHL has committed to reducing its greenhouse gas emissions by 2030 in accordance with SBTi standards and the Paris Agreement, which aims to limit global warming to 1.5 °C. The company proposes to decrease emissions from 40 million tones of CO_2e in 2021 to below 29 million tones by 2030, encompassing scopes 1, 2, and 3. This entails a 42% reduction in scopes 1 and 2, and a 25% reduction in scope 3, both by 2030 compared to 2021 [24].

For the process of measuring the indicators, we present Fig. 4, with practical examples of the measurement and monitoring stage in the context of the MRV method for freight logistics. Different types of data collected are presented, such as fuel consumption, mileage travel, type of mode used and the associated emission factor.

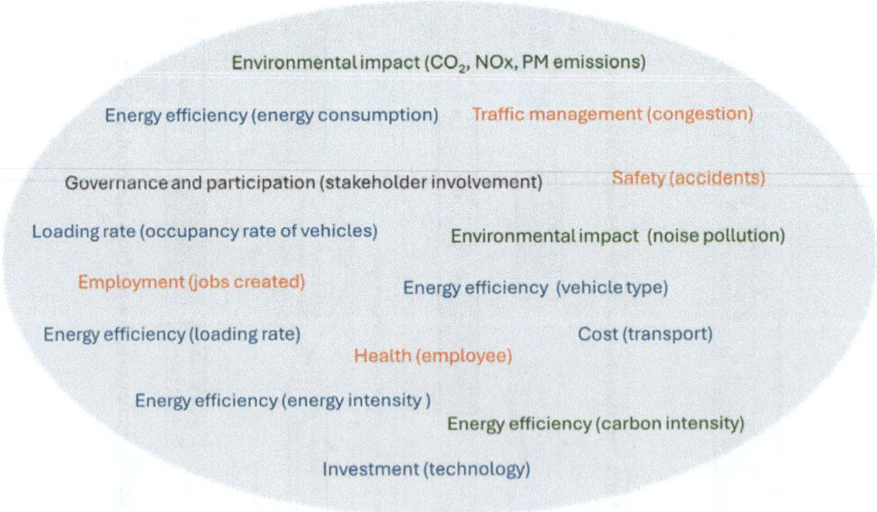

Fig. 3 Examples of indicators applied at UFT. *Source* Own elaboration (2025)

This information is essential for accurately quantifying the greenhouse gas (GHG) emissions associated with logistics operations, enabling continuous monitoring of environmental performance, the identification of opportunities to reduce emissions and the generation of more robust and reliable reports.

About monitoring, the evaluation period should encompass the entire duration of policy implementation. However, an effective monitoring framework should also include the assessment of relevant activities both before and after the policy is in place. Pre-implementation monitoring provides insights into baseline conditions, while post-implementation monitoring enables the measurement of policy outcomes and long-term impacts [49].

During the baseline verification process, the assessment should be conducted by project managers, financial backers, or specialists appointed to evaluate the initiative. The key objective is to ensure that the estimated mitigation of negative environmental impacts is both satisfactory and aligned with real-world conditions. If the assessment confirms that the projected improvements in UFT operations are adequate, the process advances to the next phase (Step 2).

However, if the results indicate inconsistencies or shortcomings, an internal report must be prepared, as outlined in the procedural framework (Fig. 2). This report should analyze the feasibility of reassessing and refining the baseline estimate. If adjustments are viable, the baseline estimation is revised accordingly. Conversely, if the reassessment is deemed unfeasible, the procedure is discontinued.

3.2 Step 2—Selection of Best Practices for UFT/Operation Monitoring

In this step, the best practices to be deployed in UFT is selected, considering the reality of enterprise's operation. The person responsible, hired or allocated to evaluate the project should select the priority actions and/or more indicated to solve the problem identified in the previous step for UFT system in which they will be implemented. With the goal to subsidize these actions' definition and select, it is suggested the use of technology maturity methods such as SLMM and TRL, SWOT analysis and others, due to its ability to model decision-making problems involving criteria Qualitative or subjective assessments by different points of view.

After best practices selection as an example presented in Fig. 5, its implementation should be monitored, in order to certify the project deadline, budget and scope, as well as to verify whether action implementation is in accordance with planning previously established by enterprise strategic level. If the action implementation is not adequate, an audit should be promoted, where non-conformities will be investigated with the technical staff. The person responsible for evaluating should decide on the feasibility of action implementation time and implantation cost, generating an internal report with the problems of non-conformities found, in order to promote new monitoring.

M | **Measurement / Monitoring**

What data and information collection?

Data source: Primary data (example: energy consumption l/year, average km per route, freight carried ton/year) and Secondary data (example: emission factors for fuels gCO_2 /km). Typically include socio-economic data related to employment, health, air quality, etc. Source: WRI (2016), Füssler et al. (2018) and D'Agosto et al. (2018)

How will be carried?

✓**Tool measuring:** Economic (BigMile, GreenRouter, LogEC - Logistics Emissions Calculator) environmental (GHG protocol tools, Global Logistics Emissions Council, EcoTransIT, BigMile, LogEC - Logistics Emissions Calculator, TK'Blue Agency – GHG calculator, GreenRouter), Social (TK'Blue Agency – GHG calculator). Source: GHGprotocol (2015), SFCA (2019)

✓ **Tool monitoring:** MRV tool, Clean Fleet toolkit. Source: Prieto (2013) and GIZ (2016)

When will be performance?

✓ Ex-ante and/or ex-post Data may be collected regularly while mitigation action is being implemented. In steps to develop performance indicators. Source: WRI (2016)

Who will carry?

✓**Entities Responsible:** Governments, national institutions, research organizations, project developers, civil society organizations, regional organizations, consultants. Source: WRI (2016)

Fig. 4 Key aspects of the measurement/monitoring stages identified in the literature. *Source* Own elaboration (2025)

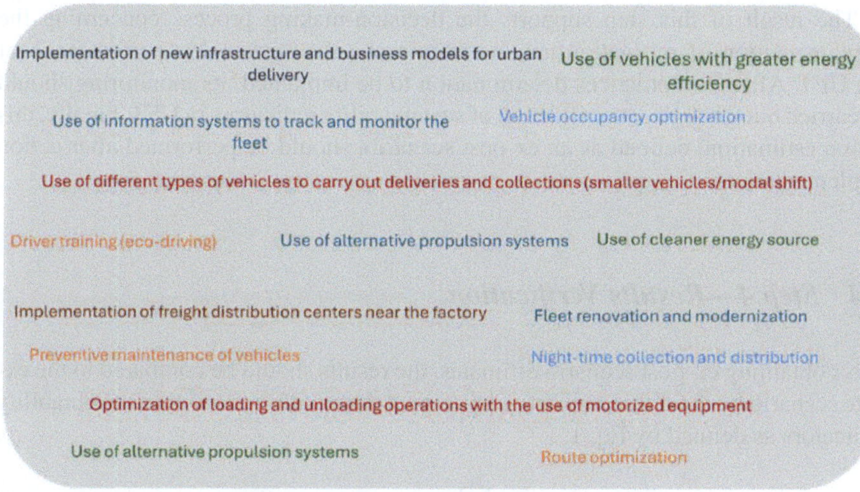

Fig. 5 Examples of best practice at UFT. *Source* Assis et al. [8, 9]

After best practices implementation, the action must be monitored in the operation phase collecting inputs and indicators of UFT operation. If there is enough data, ex-post scenario estimation is performed. Otherwise, it is necessary to draw up internal reports and to carry out new data collection.

To better understand how best practices selection and prioritization can be carried out, we cite the study by Assis et al. [8], which presents a method for identifying and prioritizing best practices in the last-mile urban delivery service, with a focus on sustainability. The research was conducted using open data from a Brazilian parcel delivery company, with the aim of optimizing its logistics operations. The method proposed in the study involves applying a prioritization matrix to select practices that promote sustainability in urban freight transport. This approach allows the company to evaluate and choose the best strategies suited to its needs and operational context.

3.3 Step 3—Ex-Post Measurement Scenario

At this stage, the best practices implanted impact on UFT operation should be estimated and reported. Mitigation index estimation can be accomplished by fulfilling targets such as NDCs, defined in the enterprise's strategic planning or meeting the specific policy requirements to the sector. If the results of the best practices are adequate, it should be directed to Step 4, in which the results are verified. However, if the results are not adequate, an internal report should be drawn up and returned to monitoring the best practices' impacts, where best practices and its impacts will be monitored again.

The result of this step supports the decision-making process concerning the implementation of mitigation actions to improve the sustainable performance of the UFT. After best practices determination to be implanted, its monitoring should be carried out through data collection of sustainability indicators in UFT. Finally, this action estimation, defined as an ex-post scenario, should be performed after action implementation.

3.4 Step 4—Results Verification

After obtaining ex-post scenario estimates, the results should be compared to the ex-ante scenario by the difference between ex-post estimation and ex-ante sustainability indicators as defined by Eq. 1.

$$\textit{Benefit Generated} = \sum I_i(Ex{-}post) - I_i(Ex{-}ante) \tag{1}$$

where "I" represents each indicator result and "i" represent economic, environmental and social aspects.

In results analysis, it is verified whether these are adequate and satisfactory. A result considered satisfactory has a positive best practices impact on sustainability indicators, besides presenting compatible results with literature. If the result is satisfactory, the result is disclosed through an external report. Otherwise, the results should be compared again, and appropriate calculations should be redone.

Thus, in Fig. 6 examples are given of who can verify and how the verification of the results obtained with the strategic objectives established and the actions implemented are aligned with the company's sustainability goals. This verification process is essential to ensure the effectiveness of the practices adopted and to provide transparent and reliable data that can be used in reports and external audits, contributing to the continuous improvement of urban logistics operations.

3.5 Step 5—External Report

The external disclosure report serves as a critical tool for transparently communicating the results of the procedure to all relevant stakeholders, including policymakers, industry representatives, regulatory agencies, financial backers, and other entities involved in the implementation of the project. Its primary objective is to ensure that the outcomes of the monitoring and evaluation process of UFT mitigation actions are clearly documented and effectively disseminated.

This report provides a comprehensive overview of the key findings, including the extent to which mitigation measures have been successfully implemented, the achieved reductions in energy and carbon intensity, and any deviations from

V	Verification

Who verified?

✓**Entities Responsible:** Can be verified through domestic laws or by the project manager or financier, technical expert review for post-2020 actions, per the Paris Agreement, is addressed at the international level, through the ICA of BURs; It can also occur at the national level, but is voluntary Source: UNFCCC(2014) and WRI (2016).

How verified?

✓**Rules and Law**: verification can be performed by prescribing domestic laws. Source: WRI (2016).

Fig. 6 Key aspects of the verification stages identified in the literature. *Source* Own elaboration (2025)

expected targets. It also outlines the methods used in data collection, monitoring, and verification, ensuring accountability and facilitating knowledge sharing among stakeholders.

Additionally, the disclosure report serves as the final step in the UFT mitigation monitoring framework, formally closing the evaluation cycle. It may include recommendations for future improvements, lessons learned, and potential policy adjustments to enhance sustainability efforts in urban freight transport. Depending on the project's scope, the report may also be shared with the broader public, contributing to transparency and fostering stakeholder engagement in the transition toward more sustainable logistics practices.

In order to better understand how reporting can be done, it is important to define which reporting model will be followed, whether GRI, SASB, CDP, among others, as well as to follow the example presented at Fig. 7.

4 Final Considerations

The proposed Selection, Measurement, Evaluation, and Reporting (SMER) method presents a comprehensive and structured framework for the identification, implementation, and validation of mitigation actions in urban freight transport (UFT). By integrating best practices and a robust set of sustainability indicators, this method enables enterprises to systematically assess the environmental and operational impacts of their initiatives. In doing so, it ensures alignment with both national and international climate commitments, including those established under frameworks such as the Paris Agreement and the United Nations Sustainable Development Goals (SDGs).

R	Reporting

What will be reportade?

✓**Informations Types:** General information (e.g. the person(s)/organisation(s) that did the assessment), objectives of assessing, steps and assessment principles, describing the policy, identifying impacts, estimating baseline emissions, estimating GHG impacts ex-ante and ex-poxt. Source (ICAT, 2018).

Who reportade?

✓**Entities Responsible:** To domestic stakeholders; to donors supporting the implementation of policies or projects. Source: WRI (2016)

When reportade?

✓**Comunications**: Biennial Reports, and/or Biennial Update Reports, and future reporting requirements yet to be determined for post-2020 contributions. WRI (2016)

Fig. 7 Key aspects of the reporting stages identified in the literature. *Source* Own elaboration (2025)

A core strength of the SMER method lies in its Monitoring, Reporting, and Verification (MRV) component, which not only supports the quantification of greenhouse gas (GHG) emissions but also establishes a transparent, standardized framework for evaluating broader, non-GHG-related co-benefits. These include improvements in logistics efficiency, urban mobility, public health through enhanced air quality, and overall sustainability performance. The method's emphasis on transparency and repeatability contributes to greater stakeholder confidence and policy alignment.

The procedural rigor of SMER—encompassing ex-ante baseline estimation, real-time monitoring of freight operations, and ex-post evaluation—demonstrates how data-driven decision-making can be effectively employed to refine sustainability strategies over time. This approach enables enterprises to identify performance gaps, implement evidence-based corrective measures, and continually adapt their mitigation efforts in response to operational realities and evolving regulatory requirements.

However, the success of the SMER method is contingent upon several critical factors. Chief among these are the availability, granularity, and accuracy of relevant data, as well as the institutional capacity and willingness of stakeholders to engage in a consistent verification process. Without these elements, the potential benefits of SMER may be compromised or unevenly realized.

Future research should therefore prioritize the exploration of emerging digital technologies—such as artificial intelligence, the Internet of Things (IoT), and blockchain—as enablers of more efficient, automated, and secure data collection and reporting processes. These technologies have the potential to significantly enhance the reliability and scalability of the SMER framework. Moreover, empirical studies and pilot projects across diverse urban contexts are essential to validate the method's applicability, adaptability, and long-term impact. Such real-world insights will not only refine the methodology itself but also inform policy development and investment decisions in sustainable urban freight systems.

References

1. Agarwal S, Kant R, Shankar R (2022) Exploring sustainability balanced scorecard for performance evaluation of humanitarian organizations. Clean Logist Supply Chain 3:100026. https://doi.org/10.1016/j.clscn.2021.100026
2. Al-Zwyalif IM (2017) Using a balanced scorecard approach to measure environmental performance: a proposed model. Int J Econ Financ 9:118–126. https://doi.org/10.5539/ijef.v9n8p118
3. Ambrosio G (2015) Guidelines: developing and implementing a sustainable urban logistics plan. https://www.eltis.org/sites/default/files/trainingmaterials/enclose_d5_2_sulp_methodology_final_version_0.pdf. Accessed 5 May 2018
4. Anand N, Quak H, Van Duin R, Tavasszy L (2012) City logistics modeling efforts: trends and gaps—a review. Procedia Soc Behav Sci 39:101–115. https://doi.org/10.1016/j.sbspro.2012.03.094
5. Andruetto C (2023) Indicators for sustainability assessment in city logistics: perspectives of society and logistic service providers
6. Asif MS, Lau H, Nakandala D, Hurriyet H (2023) Paving the way to net-zero: identifying environmental sustainability factors for business model innovation through carbon disclosure project data. Front Sustain Food Syst 7:1214490. https://doi.org/10.3389/fsufs.2023.1214490
7. Assis TFD, De Abreu VHS, Carneiro PJP, D'Agosto MDA (2021) Contribution of best practices to promote sustainable urban freight transport. Rev Geintec-Gest Inov Tecnol 11(4):5928–5949
8. Assis TFD, Abreu VHSD, Costa MGD, D'Agosto MDA (2022) Methodology for prioritizing best practices applied to the sustainable last mile—the case of a Brazilian parcel delivery service company. Sustainability 14:3812. https://doi.org/10.3390/su14073812
9. Assis TFD, Marujo LG, De Abreu VHS et al (2023) Best practices to support the transition towards sustainable logistics from the perspective of Brazilian carriers. Sustainability 15:13434. https://doi.org/10.3390/su151813434
10. Assis TFD, de Abreu VHS, Marujo LG, de Almeida D'Agosto M (2024) Best practices applied in logistic intern: from the perspective of reducing the carbon footprint. In: Carbon footprint assessments: case studies & best practices. Springer Nature Switzerland, Cham, pp 313–336
11. Assis TFD (2022) Desenvolvimento de um Método para Selecionar, Medir, Avaliar e Reportar Boas Práticas para promover a Logística Sustentável. Doctoral thesis submitted to the Graduate Program in Transportation Engineering, COPPE, Federal University of Rio de Janeiro
12. Ayadi H, Benaissa M, Hamani N, Kermad L (2024) Selecting indicators to assess the sustainability of urban freight transport using a multi-criteria analysis. Logistics 8:12. https://doi.org/10.3390/logistics8010012
13. Bandeira RAM, D'Agosto MA, Ribeiro SK, Bandeira APF, Goes GVA (2018) Fuzzy multicriteria model for evaluating sustainable urban freight transportation operations. J Clean Prod. https://doi.org/10.1016/j.jclepro.2018.02.234
14. Banister D (2008) The sustainable mobility paradigm. Transp Policy 15(2):73–80. https://doi.org/10.1016/j.tranpol.2007.10.005
15. Bartuška L, Hanzl J, Kampf R, Brlek P (2023) Indicators as a tool for assessing the level of sustainable urban freight logistics. PROMTT 35:485–499. https://doi.org/10.7307/ptt.v35i4.137
16. Bask A, Rajahonka M, Laari S et al (2016) Environmental sustainability in shipper-LSP relationships. J Clean Prod 172:2986–2998. https://doi.org/10.1016/j.jclepro.2017.11.112
17. Blanco G, de Coninck H, Agbemabiese L, Mbaye Diagne EH, Diaz Anadon L, Lim YS, Pengue WA, Sagar AD, Sugiyama T, Tanaka K, Verdolini E, Witajewski-Baltvilks J (2022) Innovation,

technology development and transfer. In: Shukla PR, Skea J, Slade R, Al Khourdajie A, van Diemen R, McCollum D, Pathak M, Some S, Vyas P, Fradera R, Belkacemi M, Hasija A, Lisboa G, Luz S, Malley J (eds) IPCC, 2022: climate change 2022: mitigation of climate change. Contribution of working group III to the sixth assessment report of the Intergovernmental Panel on Climate Change. Cambridge University Press, Cambridge, UK, New York, NY, USA. https://doi.org/10.1017/9781009157926.018
18. Browne M, Allen J, Nemoto T, Patier D, Visser J (2012) Reducing social and environmental impacts of urban freight transport: a review of some major cities. Procedia Soc Behav Sci 39:19–33
19. Buldeo Rai H, Van Lier T, Meers D, Macharis C (2018) An indicator approach to sustainable urban freight transport. J Urbanism Int Res Placemak Urban Sustain 11:81–102. https://doi.org/10.1080/17549175.2017.1363076
20. CDP (2025) Corporate environmental action tracker methodology. https://www.cdp.net/pt/data/corporate-environmental-action-tracker/corporate-environmental-action-tracker-methodology
21. CDSB (2020) The sustainability accounting standards board (SASB), climate disclosure standards board, 1, https://www.cdsb.net/sustainability-accounting-standards-board-sasb
22. CLECAT—European Association for Forwarding, Transport, Logistics and Customs Services (2010) Logistics best practice guide: a guide to implement best practices in logistics in order to save energy and reduce the environmental impact of logistics. CLECAT, Brussels, 38 pp. https://www.clecat.org/media/sr004osust091104clecatbpgv.1.0.pdf
23. Chilik I, Edens K, Klush K et al (2019) Assessment of sustainability maturity models for business transformation
24. DHL (2025) PPN 06/21—carbon reduction plan commitment to achieving net zero green logistics of choice. This carbon reduction plan captures DHL Supply Chain Limited and McGregor Cory Limited, together referred to as 'DHL Supply Chain UK'
25. Daghfous A, Zoubi T (2017) An auditing framework for knowledge-enabled supply chain management: implications for sustainability. https://doi.org/10.3390/su9050791
26. D'Agosto MA, Gonçalves DNS, Almeida IRPL, Bandeira RAM, Freitas RR, Assis TF (2018) Estudo de bases técnicas. Aplicação de metodologias nas cidades piloto. https://eemobilidadeurbana.com.br/wp-content/uploads/2018/11/Estudo-EEMU-3-Aplicac%CC%A7a%CC%83o-de-metodologias-nas-cidades-piloto.pdf
27. DeWeerdt T, Dargusch P, Hill G (2022) A case study of how DHL practices carbon management. Adv Environ Eng Res 3(1):010. https://doi.org/10.21926/aeer.2201010
28. De Abreu VHS, D'Agosto MA (2023) Logística Urbana—Tendências Atuais, Eletrificação do Transporte Urbano de Carga e Impacto da Pandemia do COVID-19. Instituto Brasileiro de Transporte Sustentável (IBTS). ISBN: 978-65-992111-6-4
29. Delipinar GE, Kocaoglu B (2016) Using SCOR model to gain competitive advantage: a literature review. Procedia Soc Behav Sci 229:398–406. https://doi.org/10.1016/j.sbspro.2016.07.150
30. European Commission (2020) Sustainable urban mobility in the EU: no substantial improvement is possible without member states' commitment. https://www.eca.europa.eu/lists/ecadocuments/sr20_06/sr_sustainable_urban_mobility_en.pdf
31. Filippi F, Nuzzolo A, Comi A, Site PD (2010) Ex-ante assessment of urban freight transport policies. Procedia Soc Behav Sci 2:6332–6342
32. Füssler J, Sharma S, Bakker S, Bongardt D, German J, Gotha S, Grütter J, Herren M, Kooshian C, Martinez H, Novikova V, Schmied M, Vieweg M (2018) Reference document on transparency in the transport sector measurement, reporting and verification of greenhouse gas emissions, 2nd updated edn. https://www.changing-transport.org/wp-content/uploads/2018_Reference_Document_Transparency-in-Transport_2nd-ed.pdf
33. Fırat SÜO, Akan MÖA, Ersoy E et al (2017) A six sigma DMAIC process for supplier performance evaluation using AHP and Kano's model. Int J Bus Anal 4:37–61. https://doi.org/10.4018/IJBAN.2017040103
34. GHG Protocol (2015) The greenhouse gas protocol. https://ghgprotocol.org/sites/default/files/standards/ghg-protocol-revised.pdf

35. GIZ (2016) MRV in practice experience in Turkey with designing and implementing a system for monitoring, reporting and verification of GHG emissions. https://www.giz.de/en/downloads/MRV_in_Practice_Booklet_2017.pdf
36. GRI (2020) About GRI. Global Reporting Initiative. https://www.globalreporting.org/about-gri/
37. Giesekam J, Norman J, Garvey A, Betts-Davies S (2021) Science-based targets: on target? Sustainability 13:1657. https://doi.org/10.3390/su13041657
38. Govindan K, Shaw M, Majumdar A (2021) Social sustainability tensions in multi-tier supply chain: a systematic literature review towards conceptual framework development. J Clean Prod 279:123075. https://doi.org/10.1016/j.jclepro.2020.123075
39. Greene S, Lewis A (2016) GLEC framework for logistics emissions methodologies. Smart Freight Centre, Amsterdam, The Netherlands
40. Gruber J, Kihm A, Lenz B (2014) A new vehicle for urban freight? An ex-ante evaluation of electric cargo bikes in courier services. Res Transp Bus Manag 11:53–62. https://doi.org/10.1016/j.rtbm.2014.03.004
41. Gudmundsdottir S, Sigurjonsson TO (2024) A need for standardized approaches to manage sustainability strategically. Sustainability 16:2319. https://doi.org/10.3390/su16062319
42. Gustavsson T (2020) Drivers for reducing GHG emissions from purchased freight transports: exploring if and how science-based targets influence companies to manage scope 3 emissions from freight transports. Master's thesis (Master in Environmental Management and Policy), Department of Industrial Economics and Management, Lund University, Lund, Sweden, 74 pp. https://lup.lub.lu.se/luur/download?func=downloadFile&recordOId=9052679&fileOId=9056565
43. Hall F (2021) A global ESG reporting standard: where are we, and where are we heading? Global Counsel, July 2021, pp 1–7. https://www.global-counsel.com/insights/report/global-esg-reporting-standard-where-are-we-and-where-are-we-heading
44. Hansen EG, Schaltegger S (2016) The sustainability balanced scorecard: a systematic review of architectures. J Bus Ethics. https://doi.org/10.1007/s10551-014-2340-3
45. Helms MM, Nixon J (2010) Exploring SWOT analysis—where are we now? A review of academic research from the last decade. J Strateg Manag 3(3):215–251. https://doi.org/10.1108/17554251011064837
46. Herold DM (2018) Has carbon disclosure become more transparent in the global logistics industry? An investigation of corporate carbon disclosure strategies between 2010 and 2015. Logistics 2(3):13. https://doi.org/10.3390/logistics2030013
47. Herold DM, Lee KH (2017) The influence of the sustainability logic on carbon disclosure in the global logistics industry: the case of DHL, FDX and UPS. Sustainability 9:601. https://doi.org/10.3390/su9040601
48. Holguín-Veras J, Leal JA, Sánchez-Diaz I, Browne M, Wojtowicz J (2020) State of the art and practice of urban freight management: part I: infrastructure, vehicle-related, and traffic operations. Transp Res Part A Policy Pract 137:360–382
49. ICAT (2018) Transport pricing: guidance for assessing the greenhouse gas impacts of transport pricing policies. How to monitor indicators over time and report results. Initiative for Climate Action Transparency. https://www.climateactiontransparency.org/wp-content/uploads/2018/05/ICAT-Transport-Ch-11-12-How-to-monitor-indicators-over-time-and-report-results.pdf
50. IPCC—Intergovernmental Panel on Climate Change (2022) Transport, chap 10. In: Shukla PR, Skinner C, Slinn N et al (eds) Climate change 2022: mitigation of climate change. Contribution of working group III to the sixth assessment report of the Intergovernmental Panel on Climate Change. Cambridge University Press, Cambridge, pp 939–1120
51. Jamshidi A, Jamshidi F, Ait-Kadi D, Ramudhin A (2019) A review of priority criteria and decision-making methods applied in selection of sustainable city logistics initiatives and collaboration partners. Int J Prod Res 57:5175–5193. https://doi.org/10.1080/00207543.2018.1540892
52. Kotsifaki M, Dimitriadis N, Ketikidis PH, Missopoulos F (2007) Logistics strategic planning: current status and future prospects in Greek companies. IJRAM 7:44. https://doi.org/10.1504/IJRAM.2007.011409

53. Lambrechts W, Son-Turan S, Reis L et al (2019) Lean, green and clean? Sustainability reporting in the logistics sector. Logistics. https://doi.org/10.3390/logistics3010003
54. Mankins JC (2009) Technology readiness assessments: a retrospective. Acta Astronaut 65:1216–1223. https://doi.org/10.1016/j.actaastro.2009.03.058
55. McKinnon A (2018) Decarbonizing logistics: distributing goods in a low carbon world. Kogan Page Publishers
56. NASA (2023) Technology readiness levels. https://www.nasa.gov/directorates/somd/space-communications-navigation-program/technology-readiness-levels/
57. Nisanci DA (2021) FSB task force on climate-related financial disclosures
58. OECD (2011) Design options for international assessment and review (IAR) and international consultations and analysis (ICA). Climate change expert group papers. Climate Change Expert Group. https://doi.org/10.1787/2227779X
59. Ouariti OZ, Bennouri J (2019) Impact of sustainable development on the performance of agro-fisheries supply chain. In: International colloquium on logistics and supply chain management, LOGISTIQUA 2019, pp 1–6. https://doi.org/10.1109/LOGISTIQUA.2019.8907245
60. Prieto EC (2013) Measurement, reporting, verification tool. NAMA approach. In: Regional workshop on promoting international collaboration to facilitate preparation, submission and implementation of NAMAs—LAC, Mexico City, 12 Dec 2013. https://unfccc.int/sites/default/files/emily_giz_mrv_tool%5B1%5D.pdf
61. Quak HJ (n.d.) Urban freight transport: the challenge of sustainability. In: City distribution and urban freight transport. https://doi.org/10.4337/9780857932754.00008
62. Regueira A, Sant'anna J, Cruz J et al (2021) Os caminhos do relato ESG um panorama sobre os frameworks e standards mais utilizados sumário
63. Revi A, Satterthwaite DE, Aragón-Durand F, Corfee-Morlot J, Kiunsi RBR, Pelling M, Roberts DC, Solecki W (2014) Urban areas. In: Field CB, Barros VR, Dokken DJ, Mach KJ, Mastrandrea MD, Bilir TE, Chatterjee M, Ebi KL, Estrada YO, Genova RC, Girma B, Kissel ES, Levy AN, MacCracken S, Mastrandrea PR, White LL (eds) Climate change 2014: impacts, adaptation, and vulnerability. Part A: global and sectoral aspects. Contribution of working group II to the fifth assessment report of the Intergovernmental Panel on Climate Change. Cambridge University Press, Cambridge, United Kingdom, New York, NY, USA, pp 535–612
64. Rich D, Bhatia P, Finnegan J, Levin K, Mitra A (2014) Policy and action standard. executive summary. An accounting and reporting standard for estimating the greenhouse gas effects of policies and actions. https://www.wri.org/publication/policy-and-action-standard. Accessed 10 Jan 2019
65. Russo F, Comi A (2011) Measures for sustainable freight transportation at urban scale: expected goals and tested results in Europe. J Urban Plan Dev. https://doi.org/10.1061/(ASCE)UP.1943-5444.0000052
66. SASB (2022) SASB standards connect business and investors on the financial impacts of sustainability. Sustainability Accounting Standards Board. https://www.sasb.org.
67. SFCA (2019) GreenRouter logistics emission calculation tool is now accredited by Smart Freight Centre. https://www.smartfreightcentre.org/
68. Stanivuk T, Gvozdenović T, Žanić Mikuličić J, Lukovac V (2020) Application of six sigma model on efficient use of vehicle fleet. Symmetry 12:857. https://doi.org/10.3390/sym12050857
69. Taniguchi E, Qureshi AG, Konda K (2020) Modelling city logistics using recent innovative technologies. Transp Res Procedia. https://doi.org/10.1016/j.trpro.2020.03.157
70. UNFCCC (2014) Annual report of the Executive Board of the clean development mechanism to the Conference of the Parties serving as the meeting of the Parties to the Kyoto Protocol. https://unfccc.int/fr/node/182445
71. VRF & IIRC (2022) Key to the development of integrated reporting and integrated thinking, Value Reporting Fundations, https://www.integratedreporting.org/the-iirc-2/structure-of-the-iirc/
72. WRI (2016) MRV 101: understanding measurement, reporting, and verification of climate change mitigation. https://transparency-partnership.net/sites/default/files/mrv_101_0.pdf

73. Werner-Lewandowska K, Golinska-Dawson P (2021) Sustainable logistics management maturity—the theoretical assessment framework and empirical results from Poland. Sustainability (Switzerland) 13(9):2021. https://doi.org/10.3390/su13095102

How Can the Implementation of Best Sustainable Practices in the Urban Freight Transport Sector Help the Logistics Sector Achieve Sustainable Development Goals?

Victor Hugo Souza de Abreu, Tássia Faria de Assis, Dalton Domingues de Carvalho Neto, and Marcio de Almeida D'Agosto

Abstract Urban logistics decision-makers face significant challenges in balancing operational efficiency with sustainability in urban freight transport (UFT). The increasing demand for fast deliveries, traffic congestion, greenhouse gas (GHG) emissions, and the need to minimize environmental impacts require the adoption of innovative and integrated strategies. In this context, it is essential to identify and implement best practices that optimize activity levels, promote the transition to low-carbon transport systems, reduce energy consumption, and enhance fuel efficiency by lowering carbon intensity. This chapter presents a comprehensive literature review aimed at mapping and analyzing the potential of best practices in UFT to contribute to the Sustainable Development Goals (SDGs) of the 2030 Agenda for Sustainable Development of the United Nations Organization. The study explores various approaches, including public policies for sustainable urban logistics, technological innovations such as fleet electrification and logistics process digitalization, and collaborative solutions like freight consolidation and the use of alternative transport modes. By gathering and evaluating these practices, this study seeks to provide valuable insights for policymakers, businesses, and researchers interested in driving the sustainable transformation of the urban freight transport sector.

Keywords Urban freight transport · Best practice · Sustainability · Carbon footprint · Mitigation · Last mile · Sustainable development goals

---

V. H. S. de Abreu (✉)
Civil Engineering at the Polytechnic School of the Federal University of Rio de Janeiro, Postdoctoral Fellow in Transportation Engineering at the Alberto Luiz Coimbra Institute for Graduate Studies and Engineering Research at the Federal University of Rio de Janeiro, Rio de Janeiro, Brazil
e-mail: victor@pet.coppe.ufrj.br; vhsa@poli.ufrj.br

T. F. de Assis · D. D. de Carvalho Neto · M. de Almeida D'Agosto
Transportation Engineering at the Alberto Luiz Coimbra Institute for Graduate Studies and Engineering Research at the Federal University of Rio de Janeiro, Rio de Janeiro, Brazil

S. S. Muthu (ed.), *Energy Efficiency and Carbon Footprint Assessment*, Environmental Footprints and Eco-design of Products and Processes,
https://doi.org/10.1007/978-3-032-10909-5_4

1 Introduction

Urban logistics companies worldwide are increasingly confronted with the dual challenge of maintaining competitiveness while advancing sustainability in an environment characterized by rising demand and continuous transformation [53, 63, 67]. Within this context, Urban Freight Transport (UFT) emerges as a pivotal element in fostering more sustainable, efficient, and livable cities [20, 36]. Companies are therefore seeking to optimize UFT operations by reducing operational costs, enhancing service quality, and minimizing environmental impacts through the adoption of internal policies that promote operational efficiency and encourage technological innovation [10, 23, 35].

Despite ongoing advancements, a substantial transformation of the freight transport sector is imperative if countries are to meet the Nationally Determined Contributions (NDCs) set forth under the 2015 Paris Agreement. A rapid and large-scale decarbonization process must be achieved by 2050 to limit global warming to below 2 °C [28, 30]. This transition demands the implementation of robust governmental policies, substantial investments in sustainable infrastructure, and the provision of incentives for adopting clean and renewable technologies [11, 12].

Beyond carbon emissions, it is equally crucial to address the broader spectrum of environmental externalities associated with UFT, including air and noise pollution, urban traffic congestion, and the deterioration of road infrastructure [2, 27, 60]. As a response to these multidimensional challenges, numerous logistics companies have begun to redesign their business models by implementing individual sustainability-oriented solutions [19]. Integrating best practices within these models has proven to be an effective means of strengthening and expanding the impact of such efforts [14, 40, 62, 70].

Given the strategic importance of best practice implementation within urban logistics, this study aims to conduct a comprehensive and systematic literature review to identify, classify, and analyze best practices applied to sustainable urban freight transport. In doing so, it seeks to establish correlations between these practices and the advancement of the Sustainable Development Goals (SDGs) outlined in the United Nations 2030 Agenda. Emphasis is placed on the contributions of these practices to operational efficiency, emissions reduction, and the development of innovative urban mobility solutions.

Moreover, the research explores globally successful strategies and initiatives, offering practical insights for companies and policymakers committed to embedding sustainability into urban freight systems. The study underscores the critical importance of multi-stakeholder collaboration, involving public authorities, private enterprises, academia, and civil society, to ensure effective and widespread implementation. It concludes by offering actionable policy recommendations tailored to decision-makers and industry leaders. These include the development of modern, sustainable infrastructure; financial and regulatory incentives; workforce training and capacity-building; and aligning local and national regulatory frameworks with international sustainability goals.

2 Methodology

This study aims to carry out a literature review, based on high-impact databases such as Scopus, Web of Science, and ScienceDirect, which are widely recognized for their scope and scientific relevance. The research seeks to identify the best practices applied to the logistics sector, with an emphasis on urban freight transportation, to promote sustainability in the sector and contribute to the fulfillment of the Sustainable Development Goals (SDGs). Thus, it is hoped to promote operational efficiency, technological innovation, and the reduction of environmental and social impacts, strengthening the resilience and competitiveness of urban logistics.

This study also aims to contextualize the current state of urban freight logistics within the broader framework of global sustainability challenges. Urban areas are experiencing increasing pressure due to population growth, economic expansion, and rising demand for goods, which often leads to congestion, pollution, and inefficiencies in freight operations. By examining best practices and innovative strategies, this research intends to provide insights into how cities can reconcile economic activity with environmental stewardship and social well-being, aligning local logistics operations with international sustainability standards.

The research also addresses the role of policy and governance in promoting sustainable urban freight. Policies at municipal, regional, and national levels can significantly influence logistics practices, from zoning regulations and traffic management to incentives for low-emission vehicles and collaborative freight initiatives. Understanding the intersection of policy frameworks and practical implementation provides a basis for recommendations that can enhance compliance, efficiency, and overall impact in urban freight systems.

To ensure methodological rigour and reliability in the results, the research follows a structured approach, organized into the following steps:

- Definition of the problem and formulation of the objectives—Initially, the relevance of the topic is established, considering contemporary challenges in urban logistics and their relationship with the SDGs. The research question is formulated based on the gaps in literature, guiding the entire investigation.
- Selection of databases and definition of search criteria—Renowned scientific databases are chosen, which guarantee access to relevant and up-to-date publications. At this stage, a set of descriptors and keywords is defined, combined with Boolean operators to refine the results, as well as inclusion and exclusion criteria to ensure the quality and relevance of the selected studies.
- Data collection, filtering, and organization—The articles identified are extracted and organized into a structured database, eliminating duplicates and material that does not meet the established inclusion and qualification criteria. Bibliographic management tools are used to systematize the references and facilitate subsequent analysis.
  - Inclusion criteria:

Specific databases: only studies extracted from high-impact and recognized databases such as Scopus, Web of Science, and ScienceDirect.
Thematic relevance: focus on practices applied to the logistics sector, especially urban freight transportation, with an emphasis on sustainability.
Relationship with SDGs: studies that contribute to the fulfillment of the SDGs.
Timeliness: Although not explicitly mentioned, the reference to "relevant and up-to-date publications" suggests a preference for recent studies.
Practical application: studies that can promote operational efficiency, technological innovation, and the reduction of environmental and social impacts.

  - Qualification Criteria (Quality and Applicability):

    Methodological quality: critical analysis of the studies in terms of their methods, scientific rigor, and the validity of the results.
    Effective contribution: studies that provide key findings and best practices relevant to the promotion of sustainable urban logistics.
    Thematic relevance: direct adherence to the themes of urban logistics, sustainability, and SDGs.
    Practical impact: applicability of the results to decision-making, public policy formulation, technological innovations, or organizational changes.
    Scientific recognition: publication in reliable sources with a high impact factor.

- Critical analysis and synthesis of results—The selected studies are evaluated in depth, considering methodological aspects, main findings, and contributions to sustainable logistics. Trends, challenges, gaps, and best practices are identified, providing a comprehensive overview of the most effective strategies in the sector.
- Discussion of findings and formulation of conclusions—The results of the review are interpreted in the light of the SDGs and the specific needs of urban freight transport. The implications for public policies, technological innovations, and organizational changes are discussed, highlighting opportunities for improvement and directions for future research.

Following this methodological path, the study seeks to consolidate a robust theoretical framework that can guide decision-making by managers, policymakers, and researchers in the search for more sustainable and efficient logistics solutions.

3 Determining Best Practices Applied to the Context

According to [24, 51], best practices are applied to a wide variety of organizations and can be planned or implemented by private companies, public companies, and public–private partnerships (PPPs). They must be aligned with a defined theme or relevant issue, based on real experiences (pilot experiments, concepts, strategies) or analysis of studies, with positive effects (qualitative and quantitative) on significant indicators for urban freight transport (UFT) [4, 36].

Among the studies already carried out are initiatives such as the VREF Center of Excellence for Sustainable Urban Freight System [38], Best Urban Freight Solutions (BESTUFS), and the Smart Freight Center (SFC). The first is a North American group, while the others are European. In Brazil, the Brazilian Green Logistics Program (Programa Logística Verde Brasil—PLVB, in Portuguese) stands out, which, through the Sustainable Reference Guide, presents best practices for freight transport [54]. In this study, the best practices were identified from a literature review using the research sources mentioned above, resulting in a set of 22 practices associated with TUC.

The Best Practices (BP) identified include: BP1 (Implementation of load consolidation centers in urban areas.), BP2 (Implementation of distribution centers close to factories), BP3 (Use of auxiliary power generation equipment to reduce fossil fuel consumption), BP4 (Installation of equipment to control vehicle emissions), BP5 (Preventive vehicle maintenance), BP6 (Optimizing vehicle occupancy), BP7 (Optimizing loading and unloading operations with motorized equipment), BP8 (Route optimization), BP9 (Promoting better vehicle aerodynamics), BP10 (Collection and distribution at night), BP11 (Shifting freight transportation to cleaner modes), BP12 (Reduced travel speed), BP13 (Reducing vehicle weight), BP14 (Fleet renewal and modernization), BP15 (Eco-driving), BP16 (Use of additives to improve the energy efficiency of fuels), BP17 (Use of different types of vehicles for deliveries and collections), BP18 (Use of cleaner energy sources), BP19 (Use of low rolling resistance tires), BP20 (Use of information systems), BP21 (Use of alternative propulsion systems), BP22 (Use of more energy-efficient vehicles).

The preliminary findings of this research, derived from case studies conducted in urban environments, are presented in Table 1. This table provides a detailed overview of the key benefits and challenges associated with the implementation of best practices, highlighting both the positive outcomes and the main obstacles encountered during their application.

The analysis of the literature reviewed (Table 1) reveals a broad spectrum of best practices applicable to UFT. While this diversity reflects the complexity and multifaceted nature of urban logistics, it can also complicate the decision-making process by creating uncertainty regarding the most appropriate options and the potential consequences of misguided choices.

From an environmental standpoint, several practices stand out for their potential to mitigate greenhouse gas (GHG) emissions and atmospheric pollutants, particularly CO_2. These include the implementation of load consolidation centers in urban areas, the establishment of distribution centers near manufacturing facilities, route optimization, and modal shifts toward transport systems with lower carbon footprints. Additionally, strategies aimed at reducing energy intensity, such as fleet renewal and modernization, also contribute to environmental gains, albeit through different mechanisms.

Notably, empirical evidence supports the differentiated effectiveness of these practices. For instance, [31] demonstrate that urban distribution centers, as implemented

Table 1 Description of best practices applied to UFT

Best practices	Description	Advantage	Disadvantages
BP1	The logistics facility is located relatively close to the geographical area served by the unit	Decreased cost, delivery time, energy consumption, air pollutants (AP), and requires initial investment; GHG; Increased safety, maintenance expenses. Reliability, capacity, and flexibility	Requires initial investment; maintenance expenses
BP2	Logistics facilities are located close to the main production sites	Decrease in cost, delivery time, energy consumption, AP and GHG; increase in safety, reliability, capacity, and flexibility	Requires initial investment; maintenance expenses
BP3	The installation of equipment to help generate energy to propel the vehicle	Reduces cost, energy consumption, AP, and GHG	Requires investment to purchase equipment
BP4	Implementation of filters and catalytic converters to control AP emissions from vehicles	Decrease in AP and GHG	Increased operating costs; requires initial investment to purchase equipment
BP5	They guarantee safe and reliable performance, preventing costly breakdowns and allowing vehicles to reach their maximum service life	Reduced cost, delivery time, energy consumption, AP, and GHG; Increased safety and reliability	Requires initial investment
BP6	It consists of maximizing the use of the vehicle's capacity	Decrease cost, energy consumption, GHG, and PA; increase flexibility	Increased delivery time; requires initial investment
BP7	It consists of using motorized equipment, which can be powered by fuel or electricity, to load and unload vehicles	Reduced energy consumption with greater energy efficiency; GHG and AP originate from cleaner fuels and the weather	Increased cost; requires initial investment
BP8	Helps reduce travel time and vehicle mileage	Reduction in cost, time, energy consumption, GHG, and AP	Requires initial investment

(continued)

Table 1 (continued)

Best practices	Description	Advantage	Disadvantages
BP9	It is responsible for installing parts and equipment in the vehicle that help to modify the wind, thus reducing its resistance	It provides a possible reduction in fuel and operating costs	It requires the installation of new equipment and may not show significant results over short distances
BP10	Cargo collection and distribution operations at alternative times to those usually practiced (business hours)	Reduced cost, delivery time, energy consumption, AP, and GHG; Increased flexibility. No initial investment required	Care must be taken with noise control
BP11	It consists of prioritizing, when possible, the use of less polluting modes of transport compared to the mode of transport in current use	Reduction in cost, energy consumption, AP, and GHG	Increased time, reduced safety and reliability; requires initial investment
BP12	Reducing the speed at which the vehicle travels	The reduction results in lower fuel consumption, lower emissions of polluting gases, increased road safety, and less wear and tear on the vehicle	It increases travel time and can have an impact on the number of customers served
BP13	Replacement of parts and components with lighter materials reduces the weight of the vehicle	Reduces wear and tear on the vehicle and its components and reduces fuel consumption	Requires initial investment
BP14	Early replacement of old vehicles	Decrease in cost, energy consumption, AP, and GHG; increase in safety and reliability	Requires initial investment
BP15	It consists of establishing a continuous and periodic training program (on average every 3 months) for drivers	Decrease in cost, energy consumption, AP, and GHG; increase in safety and reliability	Requires initial investment
BP16	It consists of the use of substances which, if added to fuels, are intended to enhance their positive properties without changing their characteristics	Reduction in energy consumption, AP, and GHG	Increases cost; requires initial investment
BP17	This involves using a greater variety of vehicles to make deliveries and collections	Reduction in time, energy consumption, GHG, and AP; increase in capacity and flexibility	Increased cost; decreased safety and reliability; requires initial investment

(continued)

Table 1 (continued)

Best practices	Description	Advantage	Disadvantages
BP18	This involves the use of alternative energy sources compared to conventional energy sources, usually oil-based fuels	Reduction in energy consumption, GHG, and AP	Increases cost; requires initial investment
BP19	Based on replacing conventional tires with low rolling resistance tires	Reduces friction between the tire and the sidewalk, enabling lower fuel consumption, greater safety, increased vehicle life, and CO_2 emissions	Increases cost; requires initial investment
BP20	It consists of using information systems such as Geographic Information Systems (GIS), Intelligent Traffic Systems (ITS), Global Positioning Systems (GPS), and Telemetry Systems to track and monitor the fleet in real time	Decrease in cost, delivery time, energy consumption, AP, and GHG; increase in safety and reliability	Requires initial investment
BP21	It consists of using a vehicle with a different propulsion system from the conventional one (internal combustion engine and mechanical transmission system)	Decrease in energy consumption, AP, and GHGs	Increases cost; requires initial investment
BP22	Conventional propulsion system with incremental improvements to its (manufacturing) design	Reduction in costs, energy consumption, GHG, and AP	Requires initial investment

Source Own elaboration based on [25, 26, 39]

in Rome, Italy, may offer greater reductions in environmental externalities than initiatives solely focused on updating vehicle fleets. This suggests the importance of contextualizing BPs according to local urban dynamics and challenges. Complementarily, Lumbreras et al. [46] indicate that while fleet modernization can enhance urban mobility and lower traffic-related emissions, its impact on CO_2 reduction is relatively limited when compared to other pollutants.

Technological innovations also play a significant role in improving environmental outcomes. The use of fuel additives and the integration of emission control technologies have shown tangible benefits in reducing local and regional air pollutants [69]. Emphasize that such measures can substantially improve fuel efficiency and reduce emissions of CO, HC, NOx, and particulate matter. Similarly, [66] identify a range of emissions control devices—including catalytic converters, diesel particulate filters, and various sensors—that help curb pollution from urban freight vehicles, particularly those related to PM and NOx.

Despite these advances, the social dimension of urban freight transport remains considerably underexplored. Among the 22 best practices analyzed, only three explicitly address social impacts. This gap underscores the pressing need for case studies and policy frameworks that better capture the societal implications of UFT strategies. It is essential that public managers, private companies, and civil society actively promote a more balanced approach that integrates the three pillars of sustainability—environmental, economic, and social—while aligning these efforts with the SDGs.

Incorporating social criteria into the design, implementation, and evaluation of BPs is thus paramount. Benefits such as improved quality of life, reduced inequality, and job creation often remain overlooked, potentially due to a lack of awareness, insufficient data, or limited institutional commitment. To address this, fostering educational campaigns, promoting multisectoral dialogue, and including social indicators in public policies and urban logistics projects are vital steps. These measures not only make social impacts more visible but also help ensure that they are treated as integral elements of sustainable urban freight transport planning.

4 Correlating Best Practices with the Sustainable Development Goals

Table 2 shows the correlation between best practices in urban freight transport and the SDGs, highlighting how these actions can contribute to building more sustainable and resilient cities. Each best practice was related to the relevant SDGs, considering their potential benefits in the environmental, social, and economic context. The implementation of these best practices not only aims to reduce the environmental impacts of transport but also to optimize resources and promote equity in access to efficient logistics solutions [27]. Through proper planning and collaboration between different sectors, it is possible to maximize the positive impacts while minimizing the challenges, creating a virtuous cycle that drives sustainable development.

Table 2 offers a strategic vision of how the benefits of each best practice can be maximized, highlighting the importance of investments in innovative technologies, public–private partnerships, and regulations that encourage the adoption of sustainable solutions. It should also be noted that general awareness is needed to understand that a deeper integration of freight transport and urban sustainability strategies can be beneficial both for the efficiency of freight transport networks and for local sustainability [45].

As can be seen in Fig. 1, which complements the information presented in Table 2, the best practices applied to urban freight transport with a focus on sustainability show the potential to contribute directly to 9 of the 17 SDGs proposed by the UN's 2030 Agenda. Among the SDGs best covered by these practices, the following stand out: SDG 9 (Industry, Innovation and Infrastructure), with 16 occurrences; SDG 17

Table 2 Correlation between best practices and the sustainable development goals

Best practice (BP)	SDGs met	How to increase benefits	How to minimize challenges
BP1—Implementation of load consolidation centers in urban areas	SDG 9 (Industry, Innovation and Infrastructure), SDG 11 (Sustainable Cities And Communities), SDG 12 (Responsible Consumption and Production), and SDG 13 (Climate Action)	Encourage public–private partnerships to create these centers, integrate with sustainable modes such as cargo bikes and electric vehicles [9, 29, 33, 67]	Ensuring tax incentives and fee reductions for companies that use these centers, as well as proper urban planning to minimize conflicts with residents [23, 63]
BP2—Implementation of distribution centers close to factories	SDG 9 (Industry, Innovation and Infrastructure), SDG 11 (Sustainable Cities and Communities), SDG 12 (Responsible Consumption and Production), SDG 13 (Climate Action), and SDG 17 (Partnerships for the Goals)	Utilize big data analytics to enhance location optimization and foster the implementation of sustainable practices in manufacturing and distribution [33, 70]	Creating planned logistics zones, facilitating licensing processes, and ensuring adequate infrastructure [50]
BP3—Use of auxiliary power generation equipment to reduce fossil fuel consumption	SDG 7 (Affordable And Clean Energy), SDG 13 (Climate Action), and SDG 17 (Partnerships for the Goals)	Invest in research and innovation to improve the efficiency of auxiliary systems, and encourage their adoption with government subsidies [32]	Create accessible lines of financing and facilitate the import of more advanced technologies [23]
BP4—Installation of equipment to control vehicle emissions	SDG 3 (Good Health and Well-being), SDG 11 (Sustainable Cities And Communities), SDG 13 (Climate Action), and SDG 17 (Partnerships for the Goals)	Encourage the mandatory adoption of this equipment in urban fleets, offering incentives to carriers that adapt [26]	Reduce taxes on emission control equipment and invest in rigorous inspection to ensure compliance with regulations [28, 29]
BP5—Preventive vehicle maintenance	SDG 3 (Good Health and Well-being), SDG 9 (Industry, Innovation and Infrastructure), SDG 11 (Sustainable Cities and Communities), SDG 13 (Climate Action), and SDG 17 (Partnerships for the Goals)	Create awareness programs on the benefits of preventive maintenance, and make periodic vehicle inspections mandatory [40, 49]	Offer technical support and financial incentives for small hauliers to join maintenance programs [26]

(continued)

Table 2 (continued)

Best practice (BP)	SDGs met	How to increase benefits	How to minimize challenges
BP6—Optimizing vehicle occupancy	SDG 9 (Industry, Innovation and Infrastructure), SDG 11 (Sustainable Cities and Communities), SDG 12 (Responsible Consumption and Production), SDG 13 (Climate Action), and SDG 17 (Partnerships for the Goals)	Create digital platforms for cargo sharing between different companies, promoting greater efficiency [8, 22, 43]	Develop regulations that encourage and facilitate cargo sharing without compromising competitiveness [19]
BP7—Optimizing loading and unloading operations with motorized equipment	SDG 8 (Decent Work and Economic Growth), SDG 9 (Industry, Innovation and Infrastructure), and SDG 17 (Partnerships for the Goals)	Automate loading and unloading processes, promoting digitalization and the Internet of Things (IoT) to reduce waiting times [22, 43, 53]	Provide training for workers who operate the equipment, ensuring safety and efficiency [29]
BP8—Route optimization	SDG 8 (Decent Work and Economic Growth), SDG 9 (Industry, Innovation and Infrastructure), SDG 11 (Sustainable Cities And Communities) and SDG 12 (Responsible Consumption and Production), and SDG 13 (Climate Action)	Apply artificial intelligence and machine learning to predict congestion and optimize routes [22, 43, 53, 55]	Create public–private partnerships to share traffic data in real time [19]
BP9—Promoting better vehicle aerodynamics	SDG 7 (Affordable and Clean Energy), SDG 9 (Industry, Innovation and Infrastructure), SDG 13 (Climate Action), and SDG 17 (Partnerships for the Goals)	Implementing mechanisms that help reduce fuel consumption and, consequently, reduce operating costs [18, 42]	Install air deflectors and other modifications to reduce wind resistance during travel [18]
BP10—Collection and distribution at night	SDG 8 (Decent Work and Economic Growth), SDG 11 (Sustainable Cities And Communities), and SDG 13 (Climate Action)	Regulate night operations to reduce daytime traffic and offer tax incentives to companies that adopt this practice [69]	Implement technologies to reduce noise during night deliveries, such as electric vehicles and silent equipment [26]

(continued)

Table 2 (continued)

Best practice (BP)	SDGs met	How to increase benefits	How to minimize challenges
BP11—Shifting freight transportation to cleaner modes	SDG 9 (Industry, Innovation and Infrastructure) and SDG 13 (Climate Action)	Expand rail and waterway infrastructure, promoting incentives for its use [26, 64], including by promoting multimodal transportation [52]	Facilitate bureaucratic processes for intermodal transportation and create attractive tariffs for companies that make the transition [32]
BP12—Reduced travel speed	SDG 3 (Good Health and Well-being), SDG 7 (Affordable And Clean Energy), SDG 11 (Sustainable Cities And Communities), and SDG 13 (Climate Action)	Encourage the adoption of reduced commuting speeds [56, 58]	Develop training and guidance for drivers on average travel speed and route optimization [18]
BP13—Reducing vehicle weight	SDG 9 (Industry, Innovation and Infrastructure), SDG 12 (Responsible Consumption and Production), SDG 13 (Climate Action), and SDG 17 (Partnerships for the Goals)	Using lighter and more resistant materials in the composition of the vehicle, which increases energy efficiency [1, 57]	Implement partnerships with the manufacturers and/or distributors of the trainsets to improve energy efficiency and with research institutions to develop technology [57]
BP14—Fleet renewal and modernization	SDG 9 (Industry, Innovation and Infrastructure) and SDG 13 (Climate Action)	Create financing programs for fleets that adopt more efficient and less polluting vehicles [26, 61, 64]	Develop public policies that accelerate the phasing out of old, highly polluting vehicles [26]
BP15—Eco-driving	SDG 4 (Quality Education), SDG 8 (Decent Work and Economic Growth), and SDG 13 (Climate Action)	Implement continuous economic driving training for drivers, using gamification to stimulate learning [18]	Regulate the requirement for eco-driving training for commercial transport drivers [24]
BP16—Use of additives to improve the energy efficiency of fuels	SDG 7 (Affordable And Clean Energy), SDG 9 (Industry, Innovation and Infrastructure), and SDG 13 (Climate Action)	Encourage research to develop additives with a lower environmental impact [18]	Create standards and certifications that guarantee the quality and efficiency of the additives used [18]

(continued)

Table 2 (continued)

Best practice (BP)	SDGs met	How to increase benefits	How to minimize challenges
BP17—Use of different types of vehicles for deliveries and collections	SDG 7 (Affordable And Clean Energy), SDG 9 (Industry, Innovation and Infrastructure), SDG 11 (Sustainable Cities And Communities), and SDG 13 (Climate Action)	Encourage the use of cargo bikes, drones, and electric vehicles for small volumes [33, 61]	Adapting urban infrastructure to allow the safe and efficient circulation of these alternative modes [71]
BP18—Use of cleaner energy sources	SDG 7 (Affordable and Clean Energy), SDG 13 (Climate Action), and SDG 17 (Partnerships for the Goals)	Expanding renewable energy supply networks for transport fleets [22, 64, 72]	Reduce taxes on clean energies or the carbon footprint and create subsidies for companies that adopt renewable sources [48]
BP19—Use of low rolling resistance tires	SDG 9 (Industry, Innovation and Infrastructure), SDG 12 (Responsible Consumption and Production), SDG 13 (Climate Action), and SDG 17 (Partnerships for the Goals)	Reducing tire friction with the sidewalk results in lower fuel consumption, a reduction in polluting gases, and an increase in the useful life of the equipment [47]	Implement partnerships with tire manufacturers and/or distributors to improve energy efficiency and with research institutions to develop technology [47]
BP20—Use of information systems	SDG 9 (Industry, Innovation and Infrastructure), SDG 12 (Responsible Consumption and Production), and SDG 13 (Climate Action)	Develop integrated platforms for sharing standardized data between companies and governments [37]	Create regulations that guarantee the security and protection of the data used [26]
BP21—Use of alternative propulsion systems	SDG 7 (Affordable And Clean Energy), SDG 9 (Industry, Innovation and Infrastructure), and SDG 13 (Climate Action)	Expand the production and use of biofuels, green hydrogen, and fleet electrification [17, 21, 26, 73]	Create incentives for the import and local production of more sustainable technologies [61]
BP22—Use of more energy-efficient vehicles	SDG 7 (Affordable And Clean Energy), SDG 9 (Industry, Innovation and Infrastructure), SDG 12 (Responsible Consumption and Production), and SDG 13 (Climate Action)	Creating energy efficiency labels for commercial vehicles, promoting R&D for more efficient engines [34, 59]	Implement policies that accelerate the transition to low-emission vehicles [34]

Source Own elaboration based on [25, 26, 39]

Fig. 1 Summary of the relationship between the best practices applied to the UFT and the SDGs. *Source* Own elaboration

(Partnerships for the goals), with 10 occurrences; and SDG 13 (Climate Action), which can be directly impacted by all the best practices mapped.

It should also be noted that other SDGs can be indirectly benefited, such as:

- SDG 1 (No Poverty): Improvements in urban logistics efficiency can reduce operating costs, contributing to more affordable prices for essential goods, generating formal jobs in the transport and urban logistics sector, and providing better use and occupation of land, generating attraction to business centers.
- SDG 2 (Zero Hunger): Efficient and sustainable urban logistics facilitates the flow of perishable food and reduces losses in the supply chain, benefiting urban food supply and improving land use, which will be able to attract more employment to the region and, consequently, generating more jobs and income for the local population and reducing hunger.
- SDG 5 (Gender Equality): The introduction of sustainable practices and logistics digitalization can expand inclusive work opportunities, including for women, who are traditionally underrepresented in the transport sector, and the inclusion of

people with special needs or neurodiversity in the development of administrative activities.

- SDG 6 (Clear Water and Sanitation): Sustainable solutions can include water pollution control (such as reducing oil spills and waste management), protect urban water bodies, and reuse carwash water.
- SDG 10 (Reduced Inequalities): Facilitating access to goods and services in peripheral urban areas and promoting more skilled jobs contributes to reducing socio-economic inequalities.
- SDG 14 (Life Below Water): Reducing urban pollution (including water pollution) and waste from transportation helps protect aquatic ecosystems in coastal and urban areas and mitigating polluting gases such as SO_2 and NOx, which are responsible for the incidence of acid rain, which in addition to intensifying corrosion in some structures, can make soil and water more acidic, damaging soil fertility, water quality with increased acidity affecting the health of aquatic organisms
- SDG 15 (Life on Land): Reducing emissions, fossil fuel consumption, and disorderly occupation indirectly contributes to the conservation of urban and peri-urban terrestrial ecosystems.
- SDG 16 (Peace, Justice and Strong Institutions): Sustainable management of urban freight transport can promote greater transparency, regulation, and governance in the sector, as well as reduce social conflicts in areas with heavy traffic.

The SDG 17 is also important in this context, according to the advantages shown in Fig. 2. This SDG, entitled "Partnerships for the Goals", stands out for its cross-cutting role in articulating efforts between different sectors of society, emphasizing the importance of strengthening the means of implementation and revitalizing global partnerships for sustainable development. The relevance of this objective lies in its ability to integrate governments, the private sector, academia, and civil society in collaborative actions based on transparency, technical cooperation, technology transfer, and the mobilization of financial and institutional resources. Thus, the practices described here not only reflect the engagement of organizations with a sustainable agenda but also show the adoption of strategies that promote effective synergies between companies and their partners, increasing the positive impact on value chains and contributing significantly to tackling contemporary environmental, social, and economic challenges.

The initiatives highlighted, in line with the guidelines proposed by the PLVB, concretely demonstrate the commitment of member companies and their outsourced partners to building strategic and collaborative alliances. These actions not only make it possible to achieve the goal of zero net carbon emissions in their operations, but also drive the energy and technological transition, promoting innovation, environmental responsibility and integration of efforts to tackle climate change.

Taniguchi and Thompson [68] list three important elements for improving urban logistics practices: (1) technological innovations, which allow data to be collected accurately and can improve data analysis with more efficient software; (2) changes

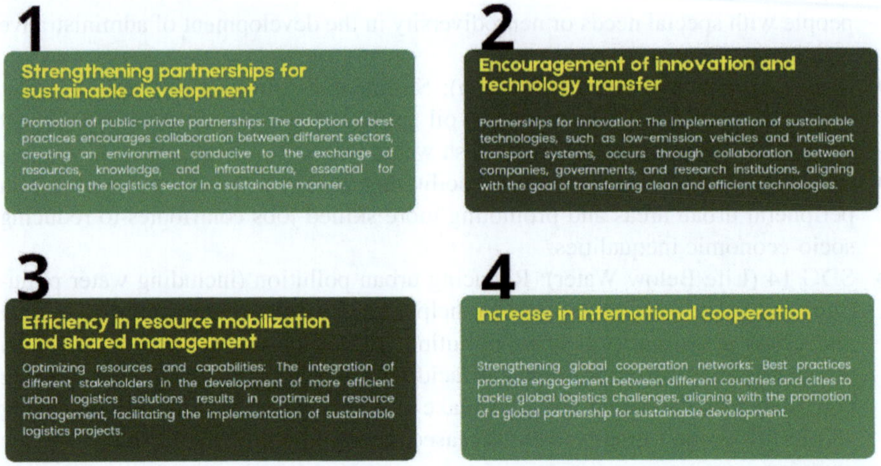

Fig. 2 Advantages of implementing best practices to achieve SDG 17. *Source* Own elaboration

in the mindset of managers to increase the sustainability of logistics activities, which is a useful way for the company to be accepted as a good partner in creating a better society and (3) public–private coordination in political decisions, in which many stakeholders are involved and decisions can generate undesirable effects for some of them.

It is essential to understand that geographical, economic, social and cultural circumstances have a direct influence on urban logistics and shape people's perceptions of the critical issues surrounding it [65]. Each city has unique characteristics that affect both the operation of freight transportation and the expectations and demands of its inhabitants. Factors such as territorial configuration, level of economic development, cultural practices, and social dynamics contribute to the complexity of the challenges faced in the urban environment [28].

Given this diversity of contexts, there is no single solution that can effectively solve all the problems related to urban freight transportation [19]. Strategies that work in certain locations may not be suitable for others, due to structural and behavioral differences [29]. For this reason, it is common to adopt a combination of initiatives, public policies, and specific measures, adjusted to local conditions [13]. This integrated approach makes it possible not only to optimize logistics flows, but also to reduce the negative impacts associated with the activity, such as congestion, pollutant emissions and conflicts with other modes of transport.

Building effective solutions for urban logistics, therefore, depends on a careful analysis of local factors and the application of an articulated set of actions [19]. Only in this way will it be possible to promote a more efficient, sustainable and resilient urban transport system, capable of meeting the needs of the population and boosting the development of cities.

At the same time, the main challenges are indicated, such as the need to adapt urban infrastructures and overcome financial and socio-economic barriers, which can be minimized by developing efficient public policies through tax incentives [7]. Thus, the implementation of best practices in urban freight transport, combined with efficient management, is a fundamental strategy for reducing GHG emissions and advancing the SDGs, promoting the transformation of cities and sustainably strengthening urban logistics.

5 The Importance of Sustainable Urban Freight Transport in City Planning

Urban freight transport occupies a critical and multifaceted role in contemporary city planning. Its relevance transcends mere logistical considerations, as it is intrinsically linked to economic efficiency, environmental sustainability, and the overall quality of life in urban areas [5, 16, 19, 44]. As cities become increasingly complex and densely populated, the efficient movement of goods becomes essential not only for supporting local economies but also for maintaining the daily functioning of urban life [61]. Figure 3 illustrates the core elements that highlight the importance of integrating urban freight transport into sustainable urban development strategies.

Considering this, [6] provide an insightful examination of how governmental bodies and planning authorities in Europe—specifically in Norway—address urban freight transport in three different cities. Through empirical case studies involving key stakeholders and real-world planning processes, the authors reveal the major

Fig. 3 The importance of UFT in city planning with a focus on sustainable development. *Source* Own elaboration

challenges and opportunities associated with integrating freight logistics into urban development strategies. Their findings emphasize that the approach adopted by urban planners significantly shapes the degree to which freight transport is effectively incorporated into broader planning frameworks.

However, as highlighted in the literature, much of the existing research in this field tends to prioritize isolated and technical solutions, such as the deployment of new technologies, the restructuring of supply chains, or the implementation of vehicle restrictions, rather than addressing the issue in a comprehensive and integrated manner. These fragmented efforts often overlook the broader institutional, social, and environmental dimensions of freight transport and fail to embed it within strategic urban governance and policy-making processes [3].

To overcome these limitations, a more holistic approach is required—one that aligns freight mobility with long-term urban development goals and governance structures. Central to this shift is the early and sustained engagement of diverse stakeholders, including public authorities, logistics operators, business actors, and civil society. Research underscores that such participatory and inclusive planning processes are crucial for the design of freight systems that are not only more efficient but also socially just and environmentally responsible [3, 15, 25]. Notably, the capacity of cities to advance in this direction is closely tied to their institutional maturity and planning coherence [41].

Additionally, [6] identify three fundamental dimensions that significantly shape how urban freight transport is integrated into planning: political anchoring, geographical scope, and temporal alignment. Political anchoring refers to the extent to which freight-related concerns are embedded in high-level policy agendas. Geographical scope involves whether freight strategies are conceived at a metropolitan scale or limited to central urban areas. Temporal alignment, in turn, concerns the synchronization of freight policies with broader urban development timelines and investment plans. These dimensions are interdependent and collectively determine the effectiveness of freight integration.

Taken together, the evidence makes clear that the successful incorporation of urban freight into city planning requires a systemic, strategic, and future-oriented approach. This entails moving beyond reactive or compartmentalized interventions and instead embedding freight issues into the very fabric of urban planning and policy-making. Ensuring robust political support, fostering stakeholder collaboration from the earliest stages, and adapting policies to the specific spatial and temporal dynamics of each city are essential steps in this process. Ultimately, enhancing urban governance and institutional capacity is indispensable for repositioning freight transport from a peripheral concern to a central component of resilient, efficient, and sustainable urban development.

6 Final Considerations

The implementation of sustainable best practices in urban freight transport plays a crucial role in tackling environmental, economic, and social challenges, contributing significantly to the achievement of the Sustainable Development Goals (SDGs). By adopting innovative strategies such as digitizing logistics, electrifying fleets, integrating multimodal transport, and establishing urban consolidation centers, it is possible to optimize freight distribution, reduce carbon emissions, and improve urban mobility.

The findings of this study highlight the potential of best practices to improve operational efficiency while minimizing negative externalities such as congestion, air pollution, and excessive energy consumption. Furthermore, by aligning sustainability principles, companies and policymakers can promote a transition towards a more resilient and environmentally responsible logistics sector. The correlation between best practices and the SDGs reinforces the importance of a holistic approach that balances economic growth with environmental preservation and social well-being.

However, despite the obvious benefits, several challenges remain in the widespread implementation of these practices. Financial constraints, lack of infrastructure, resistance to change, and regulatory barriers are some of the main obstacles that need to be addressed through coordinated efforts between governments, private companies, and research institutions. Public policies and financial incentives will be essential to facilitate the transition to greener and more efficient urban logistics models.

Furthermore, this study highlights the need for continuous research and data-driven decision-making to ensure that sustainable logistics solutions are adaptable to different urban contexts. The dynamic nature of urban freight transport requires constant innovation, stakeholder collaboration, and policy adjustments to maximize the effectiveness of sustainability initiatives.

Given the progress made in implementing Low Emission Zones (LEZ) and Zero Emission Zones (ZEZ) in various countries around the world, further studies are needed to deepen this argument. The development of these zones has proven to be effective in gradually reducing polluting gases through staggered stages of implementation, monitoring, and evaluation. Such measures not only impact urban freight transport but also extend their influence to land use and public passenger transport, contributing to social understanding of the urgency of the transition to sustainable practices. This underscores the need for future work that more comprehensively explores the effects of these policies on the different dimensions of the SDGs and their implications for the cultural and behavioral transformation of society.

References

1. Agarwal A, Mthembu L (2022) Weight optimization of heavy-duty truck chassis by optimal space fill design using light weight graphite Al GA 7–230 MMC. Mater Today: Proceed 52:1278–1287
2. Andruetto C, Stenemo E, Pernestål A (2024) Towards sustainable urban logistics: exploring the implementation of city hubs through system dynamics. Transp Res Interdisc Perspect 27:101204
3. Ballantyne EE, Lindholm M, Whiteing A (2013) A comparative study of urban freight transport planning: addressing stakeholder needs. J Transp Geogr 32:93–101
4. BESTUFS (2007) Best urban freight solutions: best practice update 2007/II, Updated handbook from Year 2003 Part II, intelligent transport systems (ITS). http://www.bestufs.net/download/BESTUFS_II/key_issuesII/BPU-2007-II_ITS.pdf. Accessed on 20 May 2025
5. Behrends S, Lindholm M, Woxenius J (2008) The impact of urban freight transport: a definition of sustainability from an actor's perspective. Transp Plan Technol 31(6):693–713
6. Bjørgen A, Ryghaug M (2022) Integration of urban freight transport in city planning: lesson learned. Transp Res Part D: Transp Environ 107:103310
7. Borowska-Stefańska M, Dulebenets MA, Sahebgharani A, Wiśniewski S, Kozieł M (2024) Evaluating low-emission-zone impacts on urban road transport system in large city. Transp Res Part D: Transp Environ 137:104503
8. Bosona T (2020) Urban freight last mile logistics—challenges and opportunities to improve sustainability: a literature review. Sustainability 12(21):8769
9. Browne M, Allen J, Woodburn A, Piotrowska M (2014) The potential for non-road modes to support environmentally friendly urban logistics. Proced Soc Behav Sci. https://doi.org/10.1016/j.sbspro.2014.10.005
10. Browne M, Allen J, Nemoto T, Patier D, Visser J (2012) Reducing social and environmental impacts of urban freight transport: a review of some major cities. Proced Soc Behav Sci. https://doi.org/10.1016/j.sbspro.2012.03.088
11. CAT Climate Action Tracker (2016) Decarbonisation series—the road ahead: how do we move to cleaner car fleets? https://climateactiontracker.org/documents/46/CAT_Decarb_Memo_Transport.pdf. Accessed on 20 May 2025
12. CAT Climate Action Tracker (2018) Decarbonisation series—the highway to Paris: safeguarding the climate by decarbonising freight transport. https://climateactiontracker.org/documents/353/CAT_20180522_DecarbSeries_FreightTransport.pdf. Accessed on 20 May 2025
13. CIVITAS WIKI (2020) Smart choices for cities. Making urban freight logistics more sustainable. Disponível em: https://civitas.eu/sites/default/files/civ_pol-an5_urban_web.pdf. Accessed on 20 May 2025
14. CLECAT (2010) Logistics best practice guide: a guide to implement best practices in logistics in order to save energy and reduce the environmental impact of logistics. 2nd edition, 2010. https://www.clecat.org/media/sr004osust091104clecatbpgv.1.0.pdf. Accessed on 20 May 2025
15. Comi A, Schiraldi MM, Buttarazzi B (2018) Smart urban freight transport: tools for planning and optimising delivery operations. Simul Model Pract Theory 88:48–61
16. Cui J, Dodson J, Hall PV (2015) Planning for urban freight transport: an overview. Transp Rev 35(5):583–598
17. da Costa MG, de Abreu VHS, de Assis TF, da Costa VX, de Almeida D'Agosto M, Santos AS (2022) Life cycle assessment and circular economy strategies for electric vehicle: a systematic review on mitigating climate change and reducing resource depletion in road transportation. Carbon Footprints Manuf Transp Ind 18:113–137
18. De Abreu VHS, Da Costa MG, Da Costa VX, De Assis TF, Santos AS, D'Agosto MDA (2022) The role of the circular economy in road transport to mitigate climate change and reduce resource depletion. Sustainability 14(14):8951
19. de Abreu VHS, D'Agosto MA (2023) Urban logistics—current trends, electrification of urban freight transport and the impact of the COVID-19 pandemic. Publisher: Brazilian Institute of Sustainable Transport (IBTS) ISBN: 978-65-992111-6-4

20. de Abreu VHS, D'Agosto MA, Marujo LG (2024) Sustainable urban transformation: the connection between electric mobility and smart grid. MIX Sustentável 10(4):31–45
21. de Abreu VHS (2024) Implementing circular economy principles in road transport: enhancing efficiency to mitigate resource depletion and carbon footprint. In: Carbon footprint assessments: case studies & best practices. Springer Nature Switzerland, Cham, pp 289-312
22. de Almeida CML, Silveira S, Jeneulis E, Fuso-Nerini F (2021) Using the sustainable development goals to evaluate possible transport policies for the city of Curitiba. Sustainability 13(21):12222
23. de Assis TF, de Abreu VHS, Marujo LG, de Almeida D'Agosto M (2024) Best practices applied in logistic intern: from the perspective of reducing the carbon footprint. In: Carbon footprint assessments: case studies & best practices. Springer Nature Switzerland, Cham, pp 313–336
24. De Assis TFD, Marujo LG, de Abreu VHS, da Costa MG, Rodrigues LM, D'Agosto MDA (2023) Best practices to support the transition towards sustainable logistics from the perspective of Brazilian carriers. Sustainability 15(18):13434
25. De Assis TF, De Abreu VHS, Carneiro PJP, D'Agosto MDA (2021) Contribution of best practices to promote sustainable urban freight transport. Revista Geintec-Gestao Inovacao e Tecnologias 11(4):5928–5949
26. de Assis TF, de Abreu VHS, D'Agosto MA (2020) Identification of good practices applied to sustainable urban freight transportation. Braz J Prod Eng 6(1):157–167
27. de Assis TF, de Abreu VHS, de Almeida DN, de Almeida Jordão E, Márcio de Almeida DA (2020) Using cleaner production as a sustainable strategy applied to logistics operations. Contrib Prod Eng Manage Sustain Energy Oper
28. de Assis TF, Monteiro TGM, de Abreu VHS, D'Agosto MDA, Santos AS (2022a) Enabling the green bonds market for sustainable transport projects based on the measure/monitoring, reporting, and verification method. In: Carbon footprints of manufacturing and transportation industries. Springer Nature Singapore, Singapore, pp 1–24
29. De Assis TFD, Abreu VHSD, Costa MGD, D'Agosto MDA (2022) Methodology for prioritizing best practices applied to the sustainable last mile—the case of a brazilian parcel delivery service company. Sustainability 14(7):3812
30. de Assis TF, Ricci LM, Monteiro TGM, de Abreu VHS, D'Agosto MDA, Santos AS (2022c) Sustainable transport indicators and mitigation actions applied to the green bond principles. In: Carbon footprints of manufacturing and transportation industries. Springer Nature Singapore, Singapore, pp 139–169
31. Filippi F, Nuzzolo A, Comi A, Site PD (2010) Ex-ante assessment of urban freight transport policies. Proced Soc Behav Sci. https://doi.org/10.1016/j.sbspro.2010.04.042
32. Fulzele V, Shankar R (2022) Improving freight transportation performance through sustainability best practices. Transp Res Part A: Policy Pract 165:285–299
33. Golinska-Dawson P, Sethanan K (2023) Sustainable urban freight for energy-efficient smart cities—systematic literature review. Energies 16(6):2617
34. Gonzalez JN, Gomez J, Vassallo JM (2022) Do urban parking restrictions and low emission zones encourage a greener mobility? Transp Res Part D: Transp Environ 107:103319
35. Greene S, Lewis A (2016) GLEC framework for logistics emissions methodologies. Smart Freight Centre. https://www.smartfreightcentre.org. Accessed on 20 May 2025
36. He Z, Haasis HD (2020) A theoretical research framework of future sustainable urban freight transport for smart cities. Sustainability 12(5):1975
37. Hofman W, Rukanova B, Tan YH, Bharosa N, Ubacht J, Rietveld E (2024) Digital infrastructures for compliance monitoring of circular economy: requirements for interoperable data spaces. In: Future of information and communication conference. Springer Nature Switzerland, Cham, pp 332–351
38. Holguín-Veras J (2015) Improving freight system performance in metropolitan area. Transforming Transportation. http://www.trb.org/Publications/Blurbs/172487.aspx. Accessed on 20 May 2025
39. ITBS (2017) Brazilian Institute for Sustainable Transportation. Sustainability Reference Guide. Available at: https://plvb.org.br/wp-content/uploads/2018/07/Guia-de-Referencias-em-Sustentabilidade-2.pdf. Accessed on April 14, 2025

40. Iwan S (2014) Adaptative approach to implementing good practices to support environmentally friendly urban freight transport management. Proced Soc Behav Sci 151:70–86
41. Kiba-Janiak M (2017) Urban freight transport in city strategic planning. Res Transp Bus Manag 24:4–16
42. Kumar BR, Nithish Varshan M, Kannan T (2019) Aerodynamic design optimization of an automobile car using computational fluid dynamics approach. Aust J Mech Eng 19(5):495–501. https://doi.org/10.1080/14484846.2019.1654963
43. Kwilinski A, Lyulyov O, Pimonenko T (2023) Environmental sustainability within attaining sustainable development goals: the role of digitalization and the transport sector. Sustainability 15(14):11282
44. Lindholm M (2010) A sustainable perspective on urban freight transport: factors affecting local authorities in the planning procedures. Proced Soc Behav Sci 2(3):6205–6216
45. Lindholm M, Behrends S (2012) Challenges in urban freight transport planning–a review in the Baltic Sea Region. J Transp Geogr 22:129–136
46. Lumbreras J, Valde's M, Borge R, Rodriguez ME (2008) Assessment of vehicle emissions projections in Madrid (Spain) from 2004 to 2012 considering several control strategies. Transp Res Part A. https://doi.org/10.1016/j.tra.2008.01.026
47. Manglano-Redondo P, Paricio-Garcia A, Lopez-Carmona MA (2025) Dynamic low-emission zones for urban mobility: a systematic review. Appl Sci 15(6):2915
48. Mardones C, Andaur C (2024) Evaluating carbon taxes in Argentina based on the demand for GHG-intensive goods and an input-output approach. Sustain Prod Consumption 46:418–429
49. Matusiewicz M (2019) Towards sustainable urban logistics: creating sustainable urban freight transport on the example of a limited accessibility zone in Gdansk. Sustainability 11(14):3879
50. McLeod S, Browne M (2023) Planning for the sustainability of freight and logistics: strategic guiding principles for regional policy. Eur J Transp Infrastruct Res 23(3):30–47
51. Mittal A, Krejci CC, Craven TJ (2018) Logistics best practices for regional food systems: a review. Sustainability. https://doi.org/10.3390/su10010168
52. Okyere S, Yang J, Adams CA (2022) Optimizing the sustainable multimodal freight transport and logistics system based on the genetic algorithm. Sustainability 14(18):11577
53. Olaleye IA, Mokogwu C, Olufemi-Phillips AQ, Adewale TT (2024) Innovative frameworks for sustainable transportation coordination to reduce carbon footprints in logistics. Int J Sci Technol Res Arch 7(02):68–75
54. Oliveira C, e D'Agosto M (2017) Reference Guide on Sustainability. Brazilian Institute for Sustainable Transportation. Rio de Janeiro. http://plvb.org.br/wp-content/uploads/2018/07/Guia-de-Referencias-em-Sustentabilidade-2.pdf. Accessed on 20 March 2025
55. Pan S, Zhou W, Piramuthu S, Giannikas V, Chen C (2021) Smart city for sustainable urban freight logistics. Int J Prod Res 59(7):2079–2089
56. Peters JF, Burguillo M, Arranz JM (2021) Low emission zones: effects on alternative-fuel vehicle uptake and fleet CO_2 emissions. Transp Res Part D: Transp Environ 95:102882
57. Petrici AV, Scutaru ML, Munteanu VM, Vlase S (2024) Use of new and light materials in automotive engineering for towing system. Appl Sci 14(17):7724
58. Piccoli A, Agresti V, Bedogni M, Lonati G, Pirovano G (2025) Modelling the air quality impacts of a Zero Emission Zone scenario in the city of Milan. Atmos Environ: X 25:100318
59. Poulhès A, Proulhac L, Malriat J (2024) Low emission zone and inequalities: non-compliant car and exposure benefit in the Paris Region. Available at SSRN 4906133
60. Rai HB, Van Lier T, Meers D, Macharis C (2017) Improving urban freight transport sustainability: policy assessment framework and case study. Res Transp Econ 64:26–35
61. Reis AC, Batista FP, Hora IMA, Abreu VHS, D'agosto MA, Oliveira MHS, Santos AS (2023) Advancements in urban freight transport: a comprehensive literature review on electric vehicles. In: 21 Congreso Chileno de Ingeniería de Transporte. Santiago. CCHIT-2023
62. Russo F, Comi A (2016) Urban freight transport planning towards green goals: synthetic environmental evidence from tested results. Sustainability 8(4):381
63. Russo F, Pellicanò DS (2019) Planning and sustainable development of urban logistics: from international goals to regional realization. WIT Trans Ecol Environ 238:59–72

64. Santos AS, de Abreu VHS, de Assis TF, Ribeiro SK, Ribeiro GM (2021) An overview on costs of shifting to sustainable road transport: a challenge for cities worldwide. Carbon Footprint Case Stud Municipal Solid Waste Manage Sustain Road Transp Carbon Sequestration 3:93–121
65. Savelsbergh M, Van Woensel T (2016) 50th anniversary invited article—city logistics: challenges and opportunities. Transp Sci 50(2):579–590. https://doi.org/10.1287/trsc.2016.0675
66. Senthil R, Arunan K, Silambarasan R, Pranesh G (2015) Effect of fuel additives on performance improvement and emission control in diesel engines. Int J Appl Eng Res 10(38):29345–29350
67. Schliwa G, Armitage R, Aziz S, Evans J, Rhoades J (2015) Sustainable city logistics—making cargo cycles viable for urban freight transport. Res Transp Bus Manage 15:50–57
68. Taniguchi E, Thompson RG (2014) City logistics: mapping the future. CRC Press
69. Tartakovsky D, Kordova-Biezuner L, Berlin E, Broday DM (2020) Air quality impacts of the low emission zone policy in Haifa. Atmos Environ 232:117472
70. Urbanowicz-Pollock PE (2019) An international analysis of sustainable urban freight transportation planning efforts
71. Versigghel J, Fransen K, Gautama S (2023) Participation, acceptability and equity aspects of urban vehicle access regulations: who benefits and who needs to adapt? Transp Res Proced 72:1193–1200
72. Mesquita AR, de Abreu VHS, Poyares CN, Santos AS (2025) Barriers to electric vehicle adoption: a framework to accelerate the transition to sustainable mobility. Sustainability 17(18):8318
73. Deabreu VHS, da Costa MG, Assis TFD, D'Agosto MDA, Rocha RS, de Paula LOD, Laissone AMS (2025) Assessing the feasibility of enzymatic biodiesel production using the 5w2h framework: a brazilian case study with distiller's corn oil. Energies 18(20):5460

Ethanol as a Vector for Decarbonization of Brazilian Coastal Shipping: An Analysis Based on the SWOT Matrix

Beatriz Correia Nadruz Chincoli and Victor Hugo Souza de Abreu

Abstract The growing pressure for low-carbon solutions in maritime transport is accelerating the search for alternative fuels, especially in Brazilian coastal shipping. This chapter evaluates ethanol as a strategic decarbonization vector for the sector, considering the international regulatory context, the maturity of the national production chain, and the technical challenges involved. The research adopts a qualitative approach, based on literature review and document analysis, complemented by a SWOT matrix (Strengths, Weaknesses, Opportunities and Threats). Results indicate that ethanol offers significant advantages in terms of greenhouse gas emissions, existing infrastructure, and alignment with public policies. However, challenges such as fleet adaptation, the absence of specific regulations, and competition with other energy alternatives limit its adoption. The study contributes to a technical reflection on sustainable pathways in commercial shipping and proposes pathways to enable the use of biofuels in Brazilian coastal shipping.

Keywords Ethanol · Decarbonization · Coastal shipping · Maritime transport · Biofuels

---

B. C. N. Chincoli
Mechanical Engineering from the Celso Suckow da Fonseca Federal Center for Technological Education (CEFET-RJ) and MBA in Environment from the Alberto Luiz Coimbra Institute for Graduate Studies and Engineering Research, Federal University of Rio de Janeiro, Rio de Janeiro, Brazil
e-mail: beatrizchincoli@gmail.com

V. H. S. de Abreu (✉)
Civil Engineering at the Polytechnic School of the Federal University of Rio de Janeiro and the Executive Postgraduate Course from the Alberto Luiz Coimbra Institute for Graduate Studies and Engineering Research, Federal University of Rio de Janeiro, Rio de Janeiro, Brazil
e-mail: victor@pet.coppe.ufrj.br; vhsa@poli.ufrj.br

S. S. Muthu (ed.), *Energy Efficiency and Carbon Footprint Assessment*, Environmental Footprints and Eco-design of Products and Processes,
https://doi.org/10.1007/978-3-032-10909-5_5

1 Introduction

Maritime transport accounts for more than 80% of global trade by volume and is responsible for around 3% of greenhouse gas (GHG) emissions worldwide. In Brazil, coastal shipping stands out, connecting ports along the coast and playing an important role in national logistics. Despite its energy efficiency compared to other modes of transport, waterborne logistics faces increasing pressure to reduce its environmental impact, especially in light of the commitments made by the International Maritime Organization (IMO) and the intensification of global climate policies [10, 14].

Vessels used in coastal shipping operate mainly on fossil fuels such as marine fuel oil (bunker) and marine diesel [1, 8], contributing significantly to the sector's national emissions. Given this scenario, there is growing interest in lower-carbon energy alternatives capable of promoting the transition to a low-carbon economy. In this context, ethanol—a renewable fuel with a consolidated production chain in Brazil—emerges as a possible solution, both for its technological maturity and its potential to reduce GHG emissions [6, 13].

The problem addressed in this study lies in the need to identify viable paths for the decarbonization of Brazilian coastal shipping, considering technological and operational solutions that reconcile environmental performance, economic viability, and regulatory compliance. The starting point is the hypothesis that ethanol, a biofuel produced on a large scale in Brazil, can be strategically incorporated into the energy matrix of coastal shipping, provided it is supported by public policies and adequate infrastructure [3, 6].

The main objective of this study is to conduct a comprehensive analysis of the potential of ethanol as a strategic vector for decarbonization in Brazilian coastal shipping. This assessment is structured around the SWOT matrix (Strengths, Weaknesses, Opportunities, and Threats), which allows for a multidimensional evaluation of the internal and external factors that influence the viability of ethanol adoption in the maritime sector.

To achieve this, the research will examine technical, economic, regulatory, and environmental dimensions, drawing upon up-to-date industry data, institutional publications, and peer-reviewed scientific literature. By integrating these perspectives, the study aims to deepen the understanding of ethanol's role in promoting a low-carbon maritime transport system and to generate evidence-based insights that can inform the development of public policies aligned with Brazil's climate targets, international commitments, and the broader transition toward sustainable logistics and energy systems.

This paper is structured as follows: Sect. 2 describes the methodology adopted. Section 3 presents the literature review, covering the context of decarbonization in maritime transport, the main energy alternatives evaluated, and the positioning of ethanol. Section 4 brings together the results obtained, with an emphasis on the SWOT analysis, which summarizes the technical, economic, and operational factors associated with its application in coastal shipping. Finally, Sect. 5 presents the final considerations and recommendations for the sector.

2 Methodology

This chapter adopts a qualitative, exploratory, and applied approach based on a review of literature and documentation, comparative analyses, and strategic evaluation. The research aims to understand the feasibility of replacing fossil fuels with ethanol in coastal shipping in Brazil, considering technical, economic, and environmental aspects, through the construction of a SWOT matrix. The initial stage consisted of reviewing technical, scientific, and institutional literature on decarbonization in maritime transport, characteristics of coastal shipping in Brazil, and the potential of ethanol as an alternative fuel, using national and international sources such as ANTAQ (National Waterway Transportation Agency), GIZ (Deutsche Gesellschaft für Internationale Zusammenarbeit), DNV (Det Norske Veritas), the BIOEN program (Bioenergy Research Program of FAPESP—São Paulo State Research Support Foundation), and the IPCC (Intergovernmental Panel on Climate Change), among others.

Next, operational data on the national coastal shipping fleet was gathered, with support from ANTAQ interactive panels and reference documents, in order to characterize the profile of ships potentially eligible for ethanol use. A comparative analysis was carried out between ethanol and the fuels currently in use, based on criteria such as emission intensity, supply feasibility, technical performance, and economic impacts. The methodology includes the construction of a SWOT matrix, whose graphical representation is shown in Fig. 1, summarizing the technical, economic, regulatory, and logistical factors associated with the adoption of ethanol in coastal shipping.

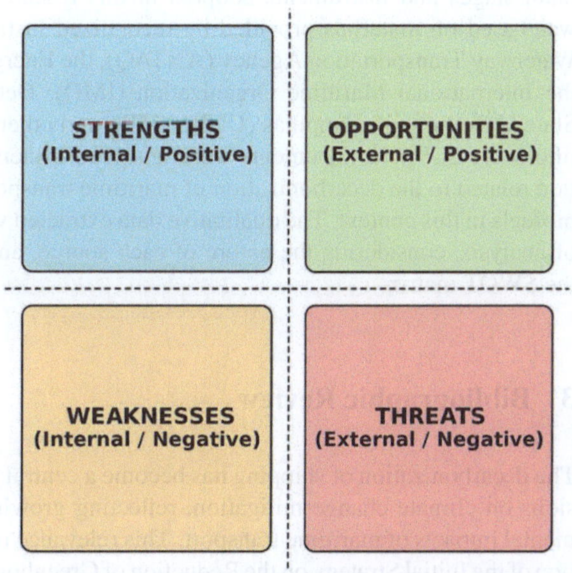

Fig. 1 Schematic representation of the SWOT matrix. *Source* Own elaboration

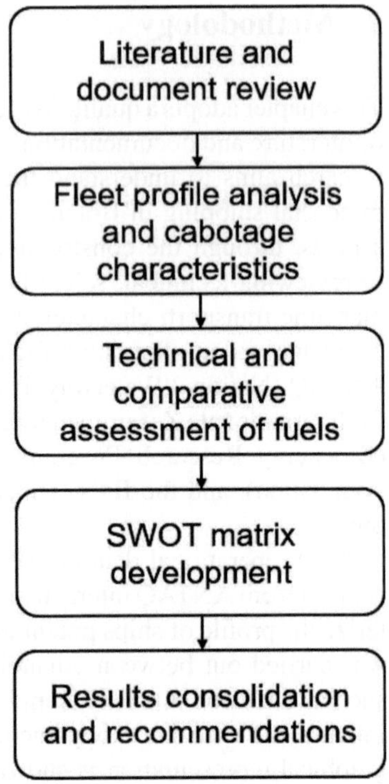

Fig. 2 Methodological stages of the research. *Source* Own elaboration

To facilitate understanding of the methodological approach, Fig. 2 presents the main stages and instruments adopted in this research. Secondary data collection was based on materials provided by recognized institutions, such as the National Waterway Transportation Agency (ANTAQ), the Energy Research Company (EPE), the International Maritime Organization (IMO), Det Norske Veritas (DNV), the State University of Campinas (UNICAMP), and other relevant technical and scientific sources. These documents were analyzed systematically, extracting information related to the decarbonization of maritime transport and the feasibility of using biofuels in this context. The qualitative data extracted were organized into categories of analysis, considering the nature of each source, and used in the construction of the SWOT matrix.

3 Bibliographic Review

The decarbonization of shipping has become a central theme in international discussions on climate change mitigation, reflecting growing concern over the environmental impacts of maritime transport. This relevance intensified following the adoption of the Initial Strategy on the Reduction of Greenhouse Gas Emissions from Ships, approved by the International Maritime Organization (IMO) in 2018. The strategy established a milestone goal of reducing total annual GHG emissions from international shipping by at least 50% by 2050, compared to 2008 levels, while pursuing efforts toward full decarbonization within this century [2, 10]. This ambitious target has catalyzed a series of global initiatives, policy adjustments, and technological innovations, placing increasing pressure on both the public and private sectors to accelerate the transition to low-carbon fuels, improve energy efficiency, and invest in sustainable maritime infrastructure [16, 17]. In this context, the search for scalable, renewable energy alternatives—such as advanced biofuels—has become not only a technological imperative but also a strategic component of climate action and sustainable development within the maritime sector.

Although maritime transport is highly energy-efficient compared to other modes, its heavy reliance on fossil fuels, such as bunker oil and marine gas oil (MGO), poses barriers to the energy transition and compliance with global climate targets [8, 14]. In Brazil, cabotage shipping plays a central role in logistical integration between coastal regions and, although governed by national regulations, tends to be impacted by international decarbonization guidelines [9]. Several alternatives are being studied to replace the most carbon-intensive fossil fuels in shipping, including liquefied natural gas (LNG), green hydrogen, ammonia, methanol, and biofuels [3, 8]. Among biofuels, ethanol stands out, offering advantages such as technological maturity, a partially consolidated logistics structure, and a strong insertion in the Brazilian energy mix [6, 7, 13].

Recent studies conducted in Brazil indicate that ethanol performs competitively in terms of avoided GHG emissions when compared to marine gas oil (MGO) and bunker fuel, especially when used in adapted compression ignition engines or in solid oxide fuel cells (SOFCs), which can directly convert ethanol into electricity with high efficiency and low pollutant emissions. Although these technological pathways are still mostly at the demonstration stage, they have already shown promising results in technical feasibility studies [8, 13]. Furthermore, the use of ethanol can generate positive externalities for the agricultural chain and for the country's climate commitments [7]. From a regulatory perspective, the IMO already recognizes biofuels as potential vectors for decarbonization, although technical barriers and regulatory gaps persist, limiting their large-scale adoption [2]. In this context, several countries—such as Singapore, Norway, the United Kingdom and Chile—have been developing national plans, financial incentives and alternative logistics routes, such as the so-called "green corridors", aimed at enabling the economic viability of sustainable fuels for navigation [3, 8]. Although Brazil has extensive experience in the production of ethanol for automotive use, its application in maritime transport requires specific studies on operational safety, technical performance, vessel adaptation, and compliance with international energy efficiency standards [12, 13]. The literature indicates that these challenges can be overcome through investments in research and development (R&D), fleet modernization programs, and public policies focused on technological innovation in maritime transportation [3, 8].

Accordingly, this study aims to address a gap that remains largely unexplored in Brazil: the assessment of ethanol as a fuel for coastal shipping vessels, considering

Table 1 Comparison of alternative fuels for maritime transport

Fuel	Source	GHG emissions	Technical challenges	Availability in Brazil	Compatibility with existing fleet
Ethanol	Biomass (sugarcane, corn, cellulose)	Up to 90% reduction (in some scenarios)	Requires engine adaptation or use in SOFC	High	Moderate (requires retrofit or dual-fuel engines)
LNG	Fossil or renewable natural gas	20–25% reduction (direct emissions)	Methane slip risks; requires cryogenic storage	Moderate	High (technology already available)
Methanol	Can be renewable (bio/green)	Up to 95% reduction (green version)	Toxic, corrosive; low energy density	Low	Moderate (dual-fuel engines)
Ammonia	Synthetic from green hydrogen	Zero carbon (if green)	Highly toxic, corrosive, lack of clear regulation	Very low	Low
(Liquid) hydrogen	Electrolysis using renewable sources	Zero carbon	Complex storage; extremely low volumetric density	Low	Very low
Marine diesel oil (MGO)	Petroleum-derived (refined)	High emissions	High cost; used in smaller vessels; pollutant	High	High
Marine fuel oil (bunker fuel)	Petroleum-derived (heavy)	High emissions	Broad compatibility; however, high climate impact	High	High

Source Adapted from, [6, 8, 12, 13]

international experiences, regulatory guidelines, and the specific characteristics of the Brazilian fleet and infrastructure. The technical comparison of alternative fuels, presented in Table 1, reinforces ethanol's position as one of the most promising options in the national context [6, 8, 13].

3.1 Profile of the Brazilian Cabotage Fleet

According to ANTAQ data, the Brazilian cabotage fleet is composed primarily of vessels designed for the transportation of liquid bulk, containers, and general cargo.

As of June 2025, the country had 211 vessels in operation dedicated to cabotage navigation, based on data gathered through the interactive *Aquarela* dashboard made available by the agency itself [4].

The fleet is distributed as follows:

- Oil and chemical tankers: account for approximately half of the operational fleet's capacity, with emphasis on activities carried out by Transpetro and other operators integrated into the national energy chain.
- Container ships: concentrate operations in the main logistics corridors of the Southeast, Northeast, and North regions.
- Bulk carriers and general cargo vessels: serve regional flows of agricultural inputs, cement, coal, and fertilizers.
- Tugboats and support vessels: complement the logistics infrastructure and ensure operations at terminals [4].

The average age of the fleet is 18 years, which highlights the need for modernization from both operational and environmental perspectives. This characteristic may be seen as an opportunity for the introduction of cleaner technologies, such as the use of alternative fuels in new projects or retrofits, particularly on short and predictable routes—typical of Brazil's coastal shipping sector [3, 8].

3.2 Sustainability and Technical Characteristics of Ethanol

One of the arguments frequently raised against the use of biofuels concerns the potential competition between energy production and food production. However, in the Brazilian context, ethanol production—mostly derived from sugarcane—has not historically caused food shortages or exerted significant pressure on staple food prices. This is due to the crops' high agricultural productivity, the country's vast cultivable land area, and the predominance of production in already established agricultural zones, especially in the Center-South region. Furthermore, practices such as crop-livestock integration and the use of byproducts from the process (such as vinasse and bagasse) reinforce the sustainability of the sugar-energy chain. Specialized literature classifies Brazilian first-generation ethanol as one of the most sustainable in the world, both from an energy and environmental standpoint [6, 13].

The incorporation of ethanol into the energy mix of coastal shipping could generate positive economic effects for Brazil, especially by strengthening the sugar and ethanol industry, stimulating regional production chains, and creating jobs in inland areas of the country. Diversifying ethanol demand also tends to increase the sector's resilience to the seasonality of the automotive market and reduce its vulnerability to international oil price fluctuations. Furthermore, the use of a low-carbon intensity fuel contributes to enhancing the country's environmental image and can open up opportunities for climate financing mechanisms and carbon credits, benefiting both agribusiness and the national shipping sector [3, 7].

In addition to greenhouse gas emissions, the choice of fuel for use in maritime transport also involves a range of technical parameters, such as energy density, flash point, and physical state. These factors have a direct impact on operational safety and logistical feasibility. Table 2 provides a comparative overview of these aspects among the main marine fuels.

The analysis of technical parameters reinforces the practical implications of fuel selection for navigation, particularly regarding the volume and weight required on board. For example, considering mass-based energy density (MJ/kg), marine diesel provides 42.7 MJ/kg, whereas ethanol delivers 26.8 MJ/kg [8, 13]. This means that approximately 60% more ethanol by mass would be needed to supply the same amount of energy. If the criterion is volumetric energy density (MJ/L), which directly

Table 2 Technical parameters of marine fuels

Fuel	Physical state (20 °C)	Energy density (MJ/kg)	Flash point (°C)	Safety/Handling notes
Ethanol	Líquid	26.8	13	Flammable; requires careful storage and ventilation
Diesel marítimo (MGO)	Líquid	42.7	> 60	Lower flammability risk; current standard on smaller vessels; easy handling
Fuel Oil (Bunker)	Viscous liquid	42.7	> 60	Requires heating for pumping; produces residues and has high viscosity
Liquefied natural gas (LNG)	Cryogenic liquid	50	< − 100	Requires cryogenic tanks; risk of leakage and explosion
Methanol	Líquid	20	11	Toxic, flammable; requires double containment systems
Ammonia	Liquefied gas under pressure	18.6	− 33	Toxic; flammable under certain conditions
(Liquid) hydrogen	Cryogenic liquid	120	< − 250	Extremely flammable; requires intense containment and refrigeration

Source Adapted from, [3, 6, 8, 12, 13]

impacts tank storage space, ethanol also presents a lower value (21.1 MJ/L versus 35.9 MJ/L for marine diesel), requiring about 70% more volume to deliver the same energy output [12].

Despite this technical limitation, adequate logistical planning and the existence of a well-established distribution infrastructure within Brazil make its application viable on short-haul, highly predictable routes—typical characteristics of Brazilian cabotage [6]. Moreover, the environmental and strategic advantages of ethanol, such as its low carbon intensity, renewable origin, and alignment with climate and agricultural public policies, outweigh the operational challenges, especially when compared to fossil fuels. Therefore, its lower energy density does not represent an insurmountable barrier, but rather a manageable variable in a scenario that demands sustainable alternatives with real application feasibility.

From an economic competitiveness perspective, alternative fuels still have higher costs compared to fossil fuels. Ethanol, for example, may cost between 1.8 and 2.5 times more per megajoule (MJ) of delivered energy than marine diesel oil (MGO) or bunker fuel (VLSFO), according to estimates by [8]. These values vary depending on the feedstock, production scale, and contractual conditions. However, this cost gap can be mitigated through appropriate public policies, such as tax relief, subsidies, and carbon pricing mechanisms. Furthermore, the adoption of ethanol may yield indirect benefits to the shipping sector, such as reduced environmental penalties, improved access to markets with stringent climate requirements, and competitive advantages in international trade [7, 8].

In addition to technical and economic factors, the international recognition of the sustainability of ethanol produced in Brazil stands out. According to BIOEN [6], the country offers exceptional agro-environmental conditions for ethanol production with high energy efficiency and low carbon intensity. Although GHG emissions comparison studies often use gasoline as a reference, there is evidence that sugarcane ethanol also has considerably lower emissions than conventional marine fuels, such as bunker oil and marine diesel (MGO), especially when considering the fuel's full life cycle. The use of agro-industrial byproducts contributes to the circularity of the production process, and the predominant use of already cultivated areas—combined with the potential use of degraded land—reinforces the idea that the sector can expand without significant impacts on food security or biodiversity [6].

As illustrated in Fig. 3, first-generation ethanol exhibits one of the lowest average GHG emissions per unit of energy among the marine fuels analyzed, reinforcing its role as a viable biofuel for coastal shipping. The comparison among fuels is based on emission factors per unit of energy established by international guidelines, such as those of the Intergovernmental Panel on Climate Change [11], which are widely used in national inventories as well as by regulatory and certification bodies. The data reflect estimates based on the full life cycle of each fuel, including production, transportation, and end use.

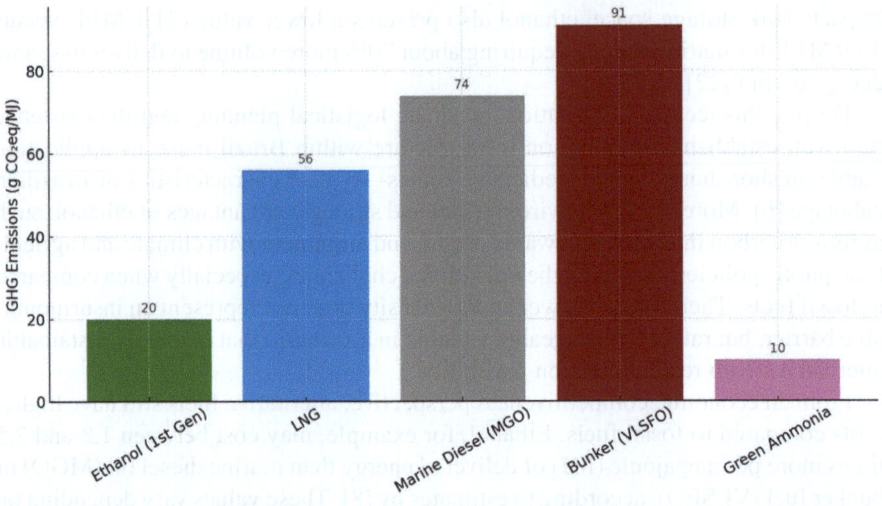

Fig. 3 Chart illustrating the comparison of average GHG emissions by type of marine fuel. *Source* Own elaboration based on [6, 8, 11]

4 Results, Challenges, and Perspectives: Ethanol in the Cabotage Energy Transition

This research aims to evaluate the potential of ethanol as a driver for the decarbonization of coastal shipping in Brazil. Based on a systematization of technical, regulatory, and scientific documents, it was possible to construct a critical assessment of this energy alternative in light of the climate and operational challenges of the maritime sector.

The results obtained from the literature review and document analysis indicate that ethanol presents promising characteristics for the decarbonization of the maritime sector, especially in the Brazilian context, where its production is well-established and its distribution infrastructure has a high degree of capillarity [6, 13]. Furthermore, global climate policies and the guidelines of the International Maritime Organization [10] have driven the development and adoption of low-carbon alternative fuels in maritime transport [2, 3].

Although Brazilian cabotage is not directly subject to regulations established by the International Maritime Organization, such as the Carbon Intensity Indicator (CII), established by the IMO [10], there is a global trend toward the incorporation of climate-related criteria into the commercial and operational policies of the maritime sector [3, 15]. Anticipating such requirements can yield not only reputational benefits but also help mitigate costs associated with emergency regulatory compliance, operational restrictions, or environmental penalties. In this context, ethanol can play a dual role: as a decarbonization driver and as a tool for managing economic and regulatory risks.

In this context, a SWOT analysis was structured, based on a review of technical, regulatory, and scientific references, with the aim of providing a comprehensive overview of the feasibility of using ethanol as an alternative energy source for the cabotage sector.

4.1 Ethanol in Cabotage: SWOT Analysis

The adoption of ethanol as an alternative fuel for cabotage shipping in Brazil was evaluated through a SWOT analysis, which considers internal strengths and weaknesses, as well as external opportunities and threats that influence its viability. The application of this tool enables the synthesis of critical factors for the success or limitations of this energy alternative in the sector. Table 3 presents the SWOT analysis, developed based on the evidence discussed throughout this study.

The SWOT analysis highlights that ethanol offers significant advantages for coastal shipping in Brazil, particularly the maturity of its production chain, international recognition of its sustainability, and alignment with global climate goals. The country has the structural conditions and expertise to lead the adoption of biofuels in the maritime sector, benefiting from its extensive logistics network, consolidated agricultural base, and the growing importance of environmental practices in international trade. However, the viability of this energy alternative depends on coordinated action among government, shipowners, and the shipbuilding industry, as well as the establishment of regulatory frameworks and economic incentives that ensure predictability, reduce transition risks, and promote technological innovation in the maritime sector.

In alignment with the United Nations 2030 Agenda for Sustainable Development, the findings of this study contribute directly to the advancement of four key Sustainable Development Goals (SDGs):

- SDG 7—Affordable and Clean Energy: The research promotes the adoption of renewable biofuels as a viable alternative to fossil fuels, fostering greater energy diversification and supporting a just energy transition toward cleaner, more accessible energy sources.
- SDG 9—Industry, Innovation and Infrastructure: By emphasizing the strategic importance of investments in logistics, port infrastructure, and technological innovation, the study reinforces the role of resilient and sustainable infrastructure in enabling low-carbon maritime transport.
- SDG 12—Responsible Consumption and Production: The proposed biofuel alternatives encourage the development of sustainable production chains and circular economy practices, contributing to more efficient resource use and reducing the environmental footprint of maritime logistics.
- SDG 13—Climate Action: By addressing the decarbonization of the maritime sector, the study contributes to global efforts to reduce greenhouse gas emissions,

Table 3 SWOT matrix: use of ethanol in Brazilian cabotage navigation

Strengths (internal)	Weaknesses (internal)
• Mature domestic production chain with an established distribution infrastructure	• Lower energy density compared to fossil fuels
• Low GHG emissions throughout the ethanol life cycle [6]	• Lower flash point, requiring additional operational safety measures
• International recognition of the sustainability of Brazilian ethanol	• Need for technological adaptation of the fleet and port infrastructure
• Alignment with international climate commitments and the sustainable development goals (SDGs)	• Lack of large-scale testing in the cabotage sector
• Potential for generating socioeconomic co-benefits in inland regions of the country	
• Possibility of adding value to the national sugar-energy chain	
Opportunities (external)	**Threats (external)**
• Brazil's regional and international leadership in biofuels and energy diplomacy	• Competition with emerging fuels that receive greater international incentives
• Promotion of public policies for decarbonization and innovation in the maritime sector	• Ethanol price volatility compared to diesel and LNG
• Development of pilot projects involving retrofitting or new vessel construction	• Alleged competition with food production, affecting public perception
• Strategic positioning of Brazil in low-carbon maritime markets	• Lack of specific national regulation for the use of ethanol as a maritime fuel in cabotage, despite international regulatory advances led by the IMO
	• Resistance from the Brazilian shipping sector to adopting new technologies in the absence of legal predictability

Source Author's elaboration based on [5–10, 12–15]

mitigate climate change impacts, and support national and international climate commitments, such as the Paris Agreement.

5 Final Considerations

This study aimed to assess the potential of ethanol as a driver for the decarbonization of coastal shipping in Brazil, using a qualitative approach based on a literature review and SWOT analysis. The research was motivated by the growing challenge of reducing greenhouse gas (GHG) emissions in the maritime transport sector, as established by the International Maritime Organization (IMO) targets, and by the need to identify solutions compatible with Brazil's productive and logistical realities. This work also contributes to the formulation of proposals aimed at the energy diversification of the maritime matrix by exploring the potential of a renewable fuel already well-established in the national context.

The analysis conducted revealed that ethanol offers a significant set of technical, economic, and environmental attributes that qualify it as a viable alternative for certain coastal shipping routes. Its main strengths include its consolidated production chain, significant reduction in GHG emissions compared to fossil fuels, and its ability to integrate with existing energy and climate policies in Brazil.

On the other hand, its adoption in the maritime sector still requires overcoming technical barriers—such as adapting the existing fleet—, the development of specific regulatory standards, and the creation of incentives to mitigate perceived risks for investors and shipping operators. In addition, competition with other low-carbon technologies and the lack of policies specifically dedicated to biofuels in shipping represent significant threats to its feasibility.

As a recommendation, it is important that Brazil develops a national program to promote the use of biofuels in maritime transport, including financing lines for fleet retrofitting, investments in research and development, and incentives for international technical cooperation. It is also suggested to carry out pilot projects and mapping priority routes to technically enable the gradual adoption of ethanol. The discussion presented in this article contributes to the development of sustainable solutions for national transportation logistics, particularly those focused on the decarbonization of cabotage, and may serve as a foundation for future studies focused on the economic and regulatory modeling of energy transition alternatives in the maritime sector. In the face of the climate emergency and the need to modernize national logistics, such initiatives represent not only a response to environmental challenges but also an opportunity to reposition Brazil as a key player in the global energy transition.

Acknowledgements I would like to express my gratitude to the Onda Azul Institute and COPPE/UFRJ for the opportunity to pursue this MBA, which marked an important milestone in my professional and academic career. I am also thankful to all the professors in the program, whose contributions were essential to expanding my knowledge and fostering critical reflection throughout this journey. A special thanks to my colleagues Michelle, Gisele, and Karla, with whom I shared learning, exchanges, and mutual support along the way. Finally, I extend my sincere appreciation to Professor Vitor for his partnership and dedication in guiding this research.

References

1. ANTAQ—National Waterway Transportation Agency (2024a) Inventory of greenhouse gas emissions from the waterway sector. Brasília, ANTAQ
2. ANTAQ—National Waterway Transportation Agency (2024b) Waterway decarbonization: concepts, practices and sustainable future. Brasília, ANTAQ
3. ANTAQ—National Waterway Transportation Agency (2024c) Review of international experience in reducing emissions in maritime transport and ports. Brasília, ANTAQ
4. ANTAQ—National Waterway Transportation Agency (2025) Information Panel on the Brazilian Cabotage fleet—aquarela system. https://aquarela.antaq.gov.br. Accessed on: June 23, 2025

5. Argus Media (2023) Brasil defende etanol como combustível marítimo. **Biofuels, Petrochemicals—Últimas notícias do mercado**, 19 out. 2023. Disponível em: https://www.argusmedia.com/pt/news-and-insights/latest-market-news/2500736-brasil-defende-etanol-como-combustivel-maritimo. Acesso em: 23 Jun 2025
6. BIOEN—Programa FAPESP de Pesquisa em Bioenergia (2025) Biocombustíveis como solução imediata e eficaz para a descarbonização do transporte. São Paulo, FAPESP
7. BRASIL. Ministério de Minas e Energia (2022) Relatório do Subcomitê de Combustíveis Marítimos—Combustíveis do Futuro. Brasília: MME. Disponível em: https://www.gov.br/mme/pt-br/programa-combustivel-do-futuro/relatorio_subcomite_combustiveis_maritimos.pdf. Acesso em: 23 Jun 2025
8. DNV (2025) Biofuels in Shipping—Current Market and Guidance on Use and Reporting. Oslo, Det Norske Veritas
9. EPE—Empresa de Pesquisa Energética (2024) Fact Sheet—Descarbonização do Transporte Aquaviário. Brasília, EPE
10. IMO—International Maritime Organization (2023) IMO Strategy on Reduction of GHG Emissions from Ships. Londres, IMO
11. IPCC—Intergovernmental Panel on Climate Change (2006) IPCC guidelines for national greenhouse gas inventories, vol 2. Energy. Disponível em: https://www.ipcc-nggip.iges.or.jp/public/2006gl/vol2.html. Acesso em: 23 Jun 2025
12. Pimenta MV, Martins MM (2021) Alternative marine fuels: relevance and viability. J Law Int Bus Acad Marit Law—MLAW 1(2):32–53
13. Souza (2022) Lucas Cesilla de. Technical-economic and environmental evaluation of marine biofuels. Dissertation (Master's)—State University of Campinas. Campinas, p 129
14. UNCTAD—United Nations Conference on Trade and Development (2023) Review of maritime transport 2023. Genebra: UNCTAD. Available at: https://unctad.org/publication/review-maritime-transport-2023. Acesso em: 23 Jun 2025
15. Waycarbon (2024) The port sector and decarbonization in maritime transport. São Paulo, WayCarbon
16. de Assis TF, Monteiro TGM, de Abreu VHS, D'Agosto MDA, Santos AS (2022a) Enabling the green bonds market for sustainable transport projects based on the measure/monitoring, reporting and verification method. In: Carbon footprints of manufacturing and transportation industries. Springer Nature Singapore, Singapore, pp 1–24
17. de Assis TF, Ricci LM, Monteiro TGM, de Abreu VHS, D'Agosto MDA, Santos AS (2022b) Sustainable transport indicators and mitigation actions applied to the green bond principles. In: Carbon footprints of manufacturing and transportation industries. Springer Nature Singapore, Singapore, pp 139–169

Critical Analysis of Biodiesel Production: Comparison of Routes and Potential of Distillers Corn Oil (DCO)

Giselle da Silva Teodoro Ferreira, Victor Hugo Souza de Abreu, and Tássia Faria de Assis

Abstract This chapter presents a critical analysis of the feasibility of biodiesel production by comparing different production routes. Special emphasis is placed on the production of biodiesel from residual oil derived from corn ethanol production, known as *Distillers Corn Oil* (DCO). The analysis begins by defining what constitutes biodiesel and highlighting its advantages over fossil-derived diesel. It then addresses the need for sustainable processes to replace fossil energy sources and explores how circular economy principles and the United Nations Sustainable Development Goals (SDGs) can be advanced through the use of biodiesel produced from DCO via an enzymatic route. The study proceeds with an evaluation of existing reaction pathways—basic homogeneous, acidic homogeneous, CaO/SiO heterogeneous, KOH heterogeneous, and enzymatic—and discusses the advantages and disadvantages of the enzymatic route in particular. Finally, a comparative life cycle assessment of fossil diesel and biodiesel is presented.

Keywords Biodiesel · Biofuel · Enzymatic route · Biodiesel production · Enzymatic catalysis

---

G. da S. Teodoro Ferreira
MBA in Environment From the Alberto Luiz Coimbra Institute for Graduate Studies and Engineering Research at the Federal University of Rio de Janeiro, Rio de Janeiro, Brazil

V. H. S. de Abreu (✉)
Civil Engineering at the Polytechnic School of the Federal University of Rio de Janeiro and Postdoctoral Fellow in Transportation Engineering at the Alberto Luiz Coimbra Institute for Graduate Studies and Engineering Research at the Federal University of Rio de Janeiro, Rio de Janeiro, Brazil
e-mail: victor@pet.coppe.ufrj.br; vhsa@poli.ufrj.br

T. F. de Assis
Transportation Engineering at the Alberto Luiz Coimbra Institute for Graduate Studies and Research in Engineering, Federal University of Rio de Janeiro, Rio de Janeiro, Brazil

S. S. Muthu (ed.), *Energy Efficiency and Carbon Footprint Assessment*, Environmental Footprints and Eco-design of Products and Processes,
https://doi.org/10.1007/978-3-032-10909-5_6

1 Introduction

Contemporary society faces a period of pronounced climate change, primarily driven by the extensive use of fossil fuels. This context demands a critical reassessment of human interactions with the environment and the prevailing patterns of natural resource exploitation [36]. The promotion of environmental preservation, the pursuit of renewable energy sources, and the implementation of sustainable practices constitute fundamental strategies for mitigating environmental impacts and advancing sustainable development [37, 38].

In this context, investing in research aimed at the development of alternative energy sources is crucial [34]. Among the promising alternatives are biofuels, defined as liquid or gaseous fuels produced predominantly from biomass [32]. Among these, biodiesel stands out due to its production versatility using various renewable sources [39].

Currently, soybeans are the main raw material used in Brazil for oil production, both for food consumption and energy generation [35]. According to the report Biodiesel in Brazil: Reflections on the Potential of Main Raw Materials [1], soybean cultivation covers approximately 47.5 million hectares. Next is corn, with around 21 million hectares under cultivation, as shown in Fig. 1. An increase in the area planted with corn is expected, mainly driven by the expansion of the ethanol industry. Considering the potential of corn and the growth of different market segments, it is likely that in the coming years, corn production will approach that of soybeans, showing greater competitiveness in some regions of the country.

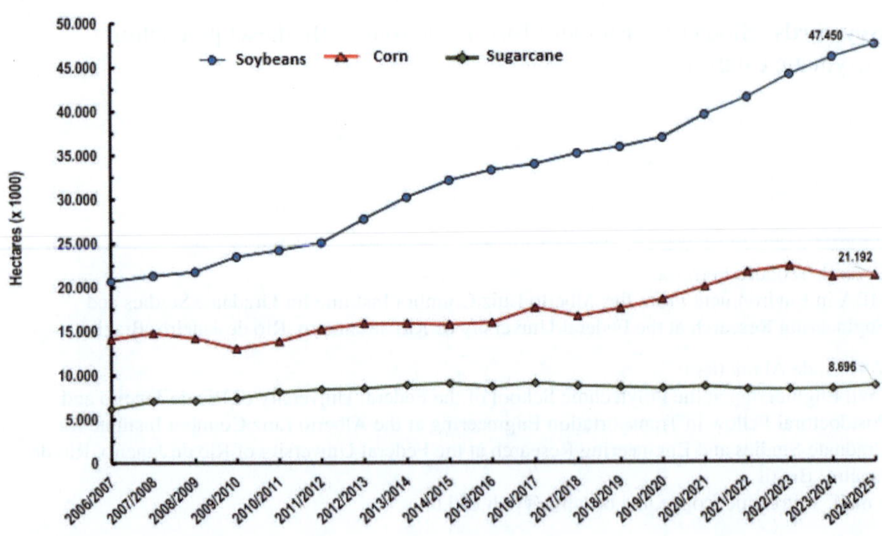

Fig. 1 Evolution of the cultivated area of the three main crops for bioenergy production in Brazil. March 2025. *Source* [1]

Currently, corn accounts for approximately 116 million tons of grain produced in Brazil, of which around 17.3 million tons are allocated to ethanol production. This volume results in the generation of 7.2 billion liters of ethanol (anhydrous and hydrated), representing 21% of the country's total ethanol output [1]. Since the 2015/2016 harvest, there has been a significant increase in ethanol production from corn, with particularly notable growth during the 2022/2023 and 2023/2024 harvests, which recorded a 33.1% rise compared to the previous period—consolidating corn ethanol as a strategic segment within the Brazilian biofuel market.

The importance of corn in the bioenergy sector extends beyond its role as animal feed. It is also a strategic crop for the production of ethanol, corn oil, and high-value by-products such as Dried Distillers Grains (DDG) and Dried Distillers Grains with Solubles (DDGS). On average, one metric ton of corn yields approximately 410 L of ethanol, 300 kg of DDG/DDGS, and 18 L of corn oil [2]. The high added value of these products has supported the rapid growth of Brazil's corn ethanol industry.

Given this context, there is a clear need to intensify scientific research aimed at the comprehensive and value-added utilization of corn, encompassing not only grains and oil but also a wide range of industrial residues generated throughout its processing chain. Among these residues, Distillers Corn Oil (DCO)—a low-nutritional-value by-product derived from the corn ethanol production process and not intended for human consumption—stands out as a particularly promising feedstock for biodiesel synthesis. Beyond offering an efficient pathway for residue valorization, the use of DCO eliminates direct competition with the food supply chain, therefore contributing to global food security and to the advancement of sustainable biofuel strategies. Additionally, its physicochemical characteristics, relative abundance, and increasing availability in biorefineries reinforce its potential as a cost-effective and environmentally advantageous raw material for renewable fuel production [40].

This chapter presents a critical analysis of biodiesel production in Brazil, with a particular focus on the comparison of different technological routes and the evaluation of the potential of DCO as a raw material. Enzymatic catalysis is explored as a promising alternative to conventional chemical processes, offering advantages such as lower environmental impact and higher reaction selectivity. The discussion also addresses the current state of Brazil's energy matrix, emphasizing the importance of diversifying feedstocks and adopting sustainable processes in the production of biofuels.

2 Methodology

This study adopts a qualitative and exploratory approach, focused on the systematization and critical analysis of scientific evidence available in the literature regarding biodiesel production through different catalytic routes, with emphasis on the application of enzymatic catalysis to residual corn oil (DCO), a by-product of corn ethanol production. This methodological choice is justified by the nature of the research object, which demands an in-depth understanding of both consolidated and emerging processes, as well as the identification of technological opportunities and challenges associated with their industrial-scale application.

Data collection was conducted through an integrative literature review, aiming to gather, describe, compare, and critically interpret the main scientific and technological advances related to biodiesel production from DCO. The following academic databases were consulted: Google Scholar, Scopus, Web of Science, and SciELO Brasil. The search employed the descriptors: biodiesel, biofuel, enzymatic route, biodiesel production, and enzymatic catalysis, used individually and in Boolean combinations. Publications in English, Portuguese, and Spanish, with a focus on studies published between 2001 and 2025, were prioritized.

The inclusion criteria for selecting studies were as follows:

- Peer-reviewed scientific publications;
- Studies focused on biodiesel production via enzymatic catalysis or those using DCO as feedstock;
- Investigations presenting quantitative data on yield, reaction time, temperature, alcohol/oil molar ratio, enzyme type and concentration, conversion efficiency, and environmental impact; and
- Studies addressing sustainability, regulatory aspects, and the economic feasibility of DCO use within the Brazilian context.

Studies with purely empirical-experimental focus lacking direct correlation with the objectives of this research, as well as reviews with an overly generic scope, were excluded. At the end of the selection process, the extracted information was organized into thematic categories, enabling a comparative analysis of catalytic routes (basic, acidic, heterogeneous, and enzymatic), the most promising operational conditions for DCO utilization, and the potential environmental and socio-economic benefits of biodiesel derived from this feedstock.

In addition to the technical-scientific evaluation of reaction parameters, the study incorporates elements of environmental and energy sustainability, based on publications applying Life Cycle Assessment (LCA) methodology and aligned with the Sustainable Development Goals (SDGs) of the United Nations 2030 Agenda. Considerations regarding the Brazilian regulatory framework, particularly those related to the production and commercialization of biofuels, were also included to support the discussion on the feasibility of integrating DCO into the national energy matrix.

It is important to highlight that this is a theoretical-analytical study and does not involve original laboratory experimentation. Its primary objective is to contribute to the advancement of technical-scientific knowledge in the field of biofuels by consolidating and confronting available evidence, identifying research gaps, and outlining perspectives for the industrial application of the enzymatic route in the valorization of agro-industrial residues such as DCO.

3 Biodiesel Production Process Using Different Technologies

This section will present the main types of catalytic routes used in biodiesel production, with an emphasis on reaction mechanisms, advantages, disadvantages and practical applications. To achieve this goal, it is important to understand the concept of biodiesel and how it is produced. Biodiesel is a biofuel intended for use in diesel engines, obtained from renewable sources such as vegetable oils and animal fats [41]. It is produced predominantly through transesterification or esterification reactions. In general, biodiesel is globally recognized as a renewable fuel derived from vegetable oils, animal fats, or waste, primarily composed of long-chain fatty acid esters (Brazil—Lei 9.478/1997, EUA—ASTM D6751, EU—EN1421). Its application is wide-ranging, covering truck engines, tractors, utility vehicles, vehicles and equipment used to generate energy.

Biodiesel consists of a mixture of fatty acid esters obtained from vegetable oils or animal fats. It can be produced through transesterification or esterification reactions. Currently, transesterification is the most widely used method, primarily because, in addition to biodiesel, it yields glycerol as a co-product of significant industrial value, which is widely applied in various sectors [42].

As mentioned above, biodiesel is primarily produced through the transesterification reaction. Transesterification is an important class of organic reactions in which one ester is transformed into another by replacing the alkoxyl group. When the original ester reacts with alcohol, the process is referred to as alcoholysis [43]. This is a reversible reaction that can occur simply by mixing the reactants; however, the use of a catalyst—classified as acidic, basic, or enzymatic—can significantly accelerate the conversion and improve the overall reaction yield.

Among the various methodologies described in the literature for biodiesel production, the transesterification of vegetable oils is currently the most widely employed. This preference is mainly due to the fact that the resulting fatty acid esters exhibit physical properties very similar to those of conventional diesel. In addition, it is a relatively straightforward process that reduces the molar mass of the compounds by approximately one third compared to triacylglycerides, while also decreasing viscosity and increasing the volatility of the final product [43].

In the transesterification of vegetable oils, a triacylglyceride reacts with alcohol in the presence of a strong base or acid, producing a mixture of fatty acid esters and glycerol (Fig. 2). The overall process consists of a sequence of three consecutive reactions, in which monoacylglycerides and diacylglycerides are formed as intermediates. For a stoichiometrically complete transesterification, a 3:1 molar ratio of alcohol to triacylglyceride is required. However, due to the reversible nature of the reaction, the transesterifying agent (alcohol) is usually added in excess, thereby increasing the ester yield and facilitating its separation from the glycerol formed [3, 4]. It is important to note that "R" in Fig. 2 represents alkyl groups containing on average 18–22 carbon atoms, and "R_1" is the alkyl group of the alcohol used as the transesterification agent.

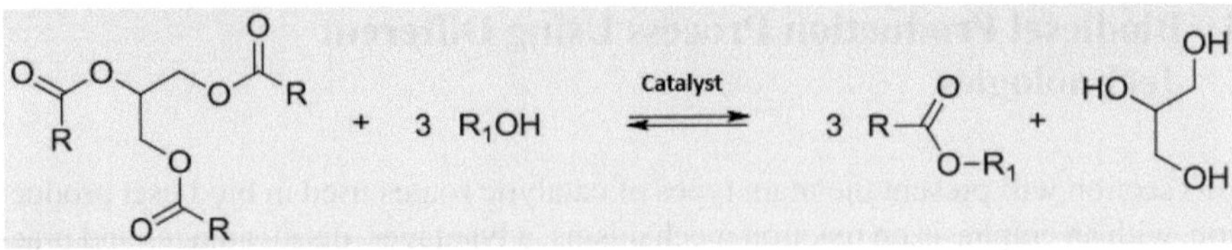

Fig. 2 Alcoholysis of vegetable oils. *Source* [4]

A wide variety of vegetable oils can be used for biodiesel production. Among the most extensively studied are soybean, sunflower, palm, almond, babassu, barley, and coconut oils [5]. The diverse fatty acid composition of these oils directly influences the physical and chemical properties of the resulting biodiesel. In addition to refined vegetable oils, residual or used cooking oils are also considered promising feedstock due to their lower cost and the potential for waste valorization. The biodiesel produced from such waste oils can achieve a quality comparable to that obtained from refined oils [43].

Regarding the transesterifying agent, the process is preferably carried out using low molecular weight alcohols such as methanol, ethanol, propanol, butanol, and amyl alcohol [6]. Among these, methanol and ethanol are the most commonly employed. Methanol is favored due to its low cost, wide availability, and advantageous physical and chemical properties, including high polarity, short carbon chain length, high reactivity with triacylglycerides, and good solubility with basic catalysts. Additionally, its use facilitates the separation of glycerol from the reaction mixture. Conversely, the use of ethanol requires greater caution, as it necessitates anhydrous ethanol and oils with low moisture content to ensure efficient glycerol separation [43].

The transesterification reaction can be conducted in acidic, basic, or enzymatic environments. Among the basic catalysts commonly used are alkali metal hydroxides, carbonates, and alkoxides, such as sodium methoxide, ethoxide, propoxide, and butoxide. Literature reports indicate that catalysts like sodium hydroxide (NaOH) and potassium hydroxide (KOH) achieve high conversion rates and good selectivity in biodiesel production [43]. However, the formation of by-products such as soaps can be minimized by employing non-ionic basic catalysts, including triethylamine, piperidine, and guanidines.

Among the most commonly used acid catalysts are sulfuric acid, sulfonic acids, and hydrochloric acid. As an alternative to conventional methods, the enzymatic conversion of vegetable oils into biodiesel presents a more environmentally sustainable approach [7]. In this process, hydrolytic enzymes, particularly lipases, serve as biocatalysts [43]. Figure 3 provides a schematic representation of the different types of catalysts employed in biodiesel production.

There are four main transesterification routes used in biodiesel production, as illustrated in the flowchart in Fig. 4 [9]. These routes primarily differ based on the type of catalyst employed or the reaction conditions applied: homogeneous catalysis, heterogeneous catalysis, enzymatic catalysis, and non-catalytic transesterification

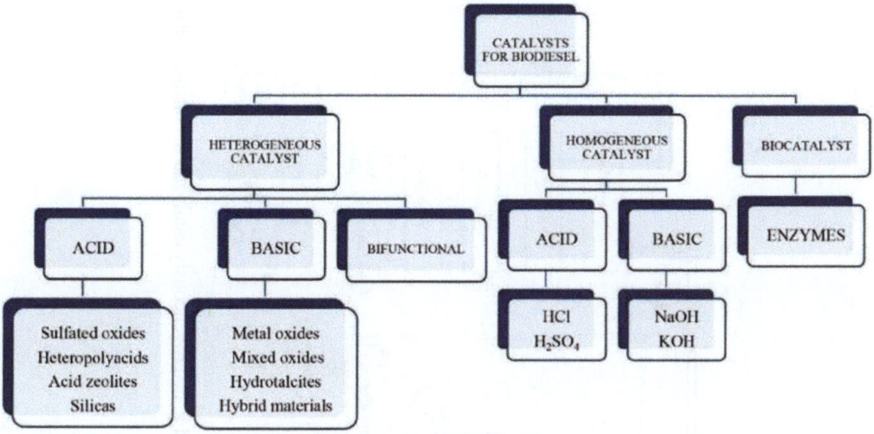

Fig. 3 Types of catalysts used in biodiesel production. *Source* [8]

using supercritical fluids. Each approach utilizes different catalytic agents—such as acids, bases, or enzymes—or relies on specific physicochemical conditions, such as elevated temperatures and pressures in the case of supercritical fluid processes.

In the following subsections, the three main catalytic routes currently employed in biodiesel production are discussed in detail: homogeneous catalysis (both acid and basic), heterogeneous catalysis, and enzymatic catalysis. Each of these routes presents distinct characteristics regarding reaction mechanisms, types of catalysts used, operational requirements (such as temperature, pressure, reaction time, and molar ratio), as well as environmental, economic, and technological implications. The analysis addresses the key parameters influencing the conversion of triacylglycerides into fatty acid esters, including reaction yield, process selectivity, by-product formation, purification requirements, and catalyst reusability.

3.1 Basic Catalysts

The most commonly used basic catalysts in biodiesel production are sodium hydroxide (NaOH) and potassium hydroxide (KOH), both of which are soluble in the reaction medium. These catalysts effectively promote the transesterification reaction but require careful handling due to their corrosive and caustic properties. Tables 1 and 2 summarize the advantages and disadvantages of using basic catalysts.

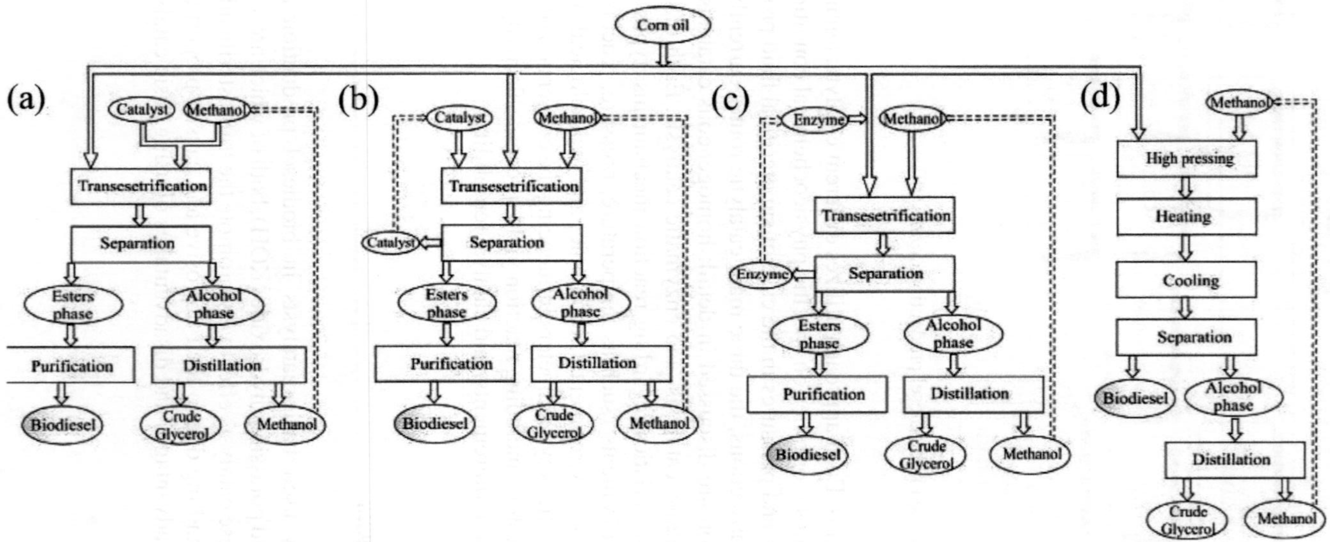

Fig. 4 Schematic presentation of the transesterification of corn oil-based feedstocks: homogeneously catalyzed **a**, heterogeneously catalyzed **b**, enzymatically catalyzed **c**, and non-catalyzed **d** reactions. *Fonte* [9]

Table 1 Advantages of using basic catalysts in biodiesel production

Aspects	Description	References
High reactivity	Reactions catalyzed by bases such as NaOH, KOH, and sodium methoxide (CH_3ONa) proceed up to 4,000 times faster than those catalyzed by acids under the same conditions	[10, 11]
Mild conditions	These base-catalyzed reactions typically operate under moderate temperature and pressure conditions	[10]
Low cost	Basic catalysts have a lower cost compared to enzymes and are widely available	[12]
Industrial efficiency	They exhibit high yields and are widely used at an industrial scale	[13]
Lower corrosiveness	Compared to acid catalysts, they cause less corrosion to industrial equipment	[18]

Table 2 Disadvantages of using basic catalysts in biodiesel production

Aspects	Description	References
Sensitivity to acidity	Oils with high free fatty acid (FFA) content generate soaps through saponification, reducing yield and complicating separation and purification	[10, 44]
Soap formations	The reaction with free fatty acids (FFA) forms soaps and water (Reaction 2.1), compromising the efficiency of the process	[14]
Catalyst consumption	The water formed binds to the catalyst, reducing its availability and requiring additional dosing, which increases the final production cost.	[18]
Necessary pretreatment	Feedstocks with high free fatty acid (FFA) content require a pretreatment step to reduce acidity	[13]
Economic impact	The additional cost of pretreatment reduces economic viability compared to the process using acid catalysts	[13]

4 Acid Catalysts

The main acid catalysts used in biodiesel production are sulfuric acid and hydrochloric acid. Tables 3 and 4 respectively present the advantages and disadvantages of using acid catalysts.

5 Enzymatic Catalysts

Lipases are enzymes that naturally catalyze both esterification and transesterification reactions. They represent one of the most important groups of biocatalysts in biotechnological applications due to their ability to operate in both aqueous and organic media, even under low-water conditions [33]. Their high application potential stems

Table 3 Advantages of using acid catalysts in biodiesel production

Aspects	Description	References
High efficiency with FFA	Acid catalysts, such as H_2SO_4 and HCl, are effective for raw materials with a high content of free fatty acids (FFA)	[15, 16]
High conversion to esters	Yield close to 99% with sulfuric acid, even in oils with high acidity	[15]
Soaps absence	No formation of soaps, which facilitates the separation and purification of biodiesel	[44]
Lower OGR costs	Processes with waste oils and fats (OGR) have shown lower total costs in acid catalysis compared to basic catalysis	[13]

Table 4 Disadvantages of using acid catalysts in biodiesel production

Aspects	Description	References
Slow reactions	The acid-catalyzed transesterification process is significantly slower than the base-catalyzed process	[15]
Harsh conditions	It requires long reaction times and temperatures above 100 °C for complete conversion	[15, 16]
Corrosiveness	The use of strong acids can cause severe corrosion of equipment, requiring more resistant and costly materials	

from their broad substrate specificity, stability across a wide range of temperatures and pH levels, resistance to organic solvents, and high chemical, positional (regioselective), and enantiomeric (enantioselective) selectivity. Currently, lipases are widely applied in various industrial sectors, including oil and fat processing, detergent and degreaser production, the food industry, synthesis of fine chemicals and pharmaceuticals, papermaking, cosmetics manufacturing, and the treatment of fatty effluents [12].

In biodiesel production, enzymatic catalysis mediated by lipases can achieve high yields; however, these yields depend on several factors, including the type of alcohol and lipid source employed, the molar ratio between reactants, reaction temperature, the presence and nature of any organic solvent, water content, and the specific origin of the lipase [12].

The use of enzymes as catalysts enables transesterification to proceed at lower temperatures (30–60 °C), thereby reducing energy consumption and improving process efficiency due to their high specificity Fantinel [17, 18]. Conversion may occur homogeneously, with lipases in solution, or heterogeneously, with immobilized enzymes. The latter offers operational advantages, as immobilization allows enzyme reuse, making the process more cost-effective and environmentally sustainable [17].

Despite the high cost of enzymes, the enzymatic route offers several advantages over the conventional chemical process [40]. These include the elimination of catalyst neutralization steps, reduced generation of waste and by-products, higher selectivity, and the ability to process diverse feedstocks in a single step. Furthermore, the environmental impact is minimized, as no toxic effluents are produced, and product recovery is facilitated [18]. Table 5 presents the advantages of using enzymes in biodiesel production, while Table 6 summarizes the disadvantages of employing enzymes as catalysts in this process.

Table 5 Advantages of using enzymes in biodiesel production

Aspects	Description	References
Less complex operation	Simpler process compared to acid and basic systems, with less risk of forming undesirable by-products	[12, 18]
Reuse by immobilization	Immobilizing the enzyme allows it to be reused in several cycles, reducing costs and increasing the stability of the process	[12]
Quality of the end product	Enzymatic catalysis favors greater purity and less contamination in the biodiesel obtained	[12, 19]
Mild operating conditions	The reaction takes place at moderate temperatures, between 20 and 60 °C, reducing energy consumption	[20]
Avoids soap formation	Even with highly acidic raw materials, saponification does not occur, making it easier to separate the product	[18]

Table 6 Disadvantages of using enzymes in biodiesel production

Aspects	Description	References
High cost of enzymes	Lipases are expensive catalysts, which limit their application on a large industrial scale	[18, 21]
Slow process	The reaction time is long and can reach 24 h for yields close to 90% by mass	[18]
Need for immobilization	Without immobilization, the enzyme can quickly lose its activity, compromising the economic viability of the process	[18, 19]
Precise operational control	Requires specific control of conditions such as temperature, pH and water content to maintain the catalytic activity of the enzyme	[20]
Complex immobilization process	The immobilization technique can involve sophisticated steps, such as chemical bonds or encapsulation, requiring specific materials and technical know-how	[18]

6 Enzymatic Route Using DCO for Biodiesel Production

The production of biodiesel from DCO has been studied as a sustainable alternative, especially in the context of utilizing by-products generated in the production of corn ethanol, such as DDGS. DCO, obtained from the distillation of DDGS, is a low-cost raw material that is widely available on the market [9]. Although corn oil has historically been considered unsuitable for biodiesel production due to its high added value and its use in human food, recent studies indicate that by-products such as corn germ and DDGS, after DCO extraction, are viable for this purpose [22].

The conversion of DCO into biodiesel can be achieved through transesterification or esterification reactions [9]. Experimental studies have shown that enzymatic transesterification of transgenic corn oil using immobilized enzymes can result in high yields of fatty acid ethyl esters (FAEE), even when some quality parameters do not fully meet the requirements of EN 14,214:2009 [23]. The characterization of DCO reveals that its composition is mainly formed by triacylglycerols, with high levels of free fatty acids (FFA), which requires the adoption of technologies suitable for its conversion, such as enzymatic catalysis [24].

Corn oil extraction from DDGS is commonly performed using methods such as solvent extraction or supercritical carbon dioxide extraction, with anhydrous ethanol being one of the most widely used solvents [25]. The ethanol extraction process can achieve yields of up to 70% [26]. However, raw DCO presents operational challenges, such as high acidity and moisture content, which makes the enzymatic route a promising alternative, as it eliminates the need for pretreatment to reduce acidity, which is required in conventional chemical processes [17, 27].

Despite its low degree of technological maturity, the enzymatic route has shown promise for DCO conversion, especially due to its ability to process oils with high FFA content without the need for prior treatment [24]. However, challenges remain in terms of costs and economic viability, which requires further studies to assess the potential of this route on a commercial scale. Future research should focus on optimizing operating parameters, improving efficiency, and reducing costs, as well as promoting integration between the corn ethanol and biodiesel production chains in order to make industrial plants economically viable [9]. In addition, it is essential to consider the properties of the biodiesel obtained, its performance in engines, the associated atmospheric emissions, and the environmental and socioeconomic impacts of its production [9].

Table 7 presents a comparison between different studies of enzymatic transesterification of corn oil using immobilized lipases. The main factors analyzed include reaction time, temperature, enzyme quantity, molar ratio of alcohol to oil, type of enzyme used, and conversion obtained.

The reaction time varies between 9 and 12 h, with the shortest time obtained using Novozym [22] while the longest time was recorded with Thermomyces lanuginosa [23]. This indicates that the type of enzyme and reaction conditions directly influence the efficiency of the process. Regarding temperature, the studies used values between 35 °C and 60 °C. The lowest temperature was applied with Thermomyces lanuginosa

Table 7 Comparison of operating parameters for biodiesel production via enzymatic catalysis (with corn oil)

Studies	Time (h)	Enzyme (% oil)	Alcohol/oil (mol/mol)	Time of enzyme	Conversion (%)
[23]	12	2.5	6:1	Thermomyces lanuginosa	98.9
[29]	4	5–15	3:1–9:1	Novozym 435 Supported	81
[22]	18	10–30	2:1–20:1	Novozym 435 Supported	91
[28]	12	0.4–3.6	1:1–9:1	Novozym 435 Supported	93

[23], while Novozym 435 was used at higher temperatures [22, 28, 29]. Temperature directly affects enzyme activity and stability, making it a critical parameter for optimization.

The amount of enzyme used in studies varies from 2.5% to 20% of the oil weight. Studies with Novozym 435 [22, 29] tend to use higher enzyme concentrations, which can contribute to greater conversion but also increases process costs. On the other hand, the molar ratio of alcohol to oil was maintained at 6:1 in all studies, indicating that this ratio is ideal for maximizing conversion without excessive alcohol waste.

The lipases Thermomyces lanuginosa and Novozym 435 were the most widely used in the studies. Novozym 435 showed greater versatility and efficiency under different reaction conditions, while Thermomyces lanuginosa presented the highest conversion (98.9%) [23]. The conversion obtained in the studies ranges from 72.9% to 98.9% [23, 29], showing that the type of enzyme and reaction parameters have a significant impact on the efficiency of the process.

In summary, enzymatic transesterification is a promising alternative for biodiesel production, especially due to its selectivity and milder reaction conditions. However, the cost of enzymes and the need to optimize reaction parameters are still challenges for its large-scale application. Future studies should focus on enzyme reuse and reducing operating costs to make the process more competitive.

7 Sustainable Development Goal

The production of biodiesel from residual oil derived from corn ethanol (DCO) is a sustainable technological alternative, aligned with several Sustainable Development Goals (SDGs) proposed by the United Nations (UN) 2030 Agenda. Table 8 shows the main SDGs covered by this approach, as well as the specific targets to which it contributes, according to the Millennium Development Goals (MDGs) [45].

Table 8 Contributions of biodiesel production from DCO to the sustainable development goals (SDGs)

SDGs	How it is handled	Related goal
SDG 2—Zero hunger	The use of industrial waste, such as DCO, which does not compete with food production, avoids pressure on food crops and promotes food security	Ensure sustainable food production systems and implement resilient agricultural practices
SDG 7—Affordable and clean energy	Replacing fossil fuels with biodiesel from agro-industrial waste, such as DCO, promotes the use of renewable and clean sources, contributing to a more sustainable energy matrix	Substantially increase the share of renewable energy in the global energy mix
SDG 8—Decent work and economic growth	The biodiesel production chain based on DCO has the potential to generate skilled jobs and promote regional development, contributing to the strengthening of the bioenergy sector	Diversify and update technologies to achieve greater economic productivity
SDG 9—Industry, innovation, and infrastructure	The adoption of the enzymatic route in biodiesel production represents a technological innovation that modernizes the biofuel sector, making industrial processes cleaner and more efficient	Upgrade infrastructure and modernize industries to make them sustainable, with more efficient use of resources
SDG 12—Responsible consumption and production	The reuse of DCO, a by-product of ethanol production, prevents waste and encourages circular economy practices by transforming waste into value-added inputs	Substantially reduce waste generation through prevention, reduction, recycling, and reuse

(continued)

Table 8 (continued)

SDGs	How it is handled	Related goal
SDG 13—Action climate	The use of biodiesel contributes to the mitigation of greenhouse gas (GHG) emissions, such as CO_2, especially when associated with life cycle assessment and emissions calculation based on IPCC methodology	Integrate climate change measures into national policies, strategies, and planning
SDG 17—Partnerships for the goals	One example is the project "Economic, Financial, and Environmental Pre-Feasibility Study of the Scale of Biodiesel Production from Residual Oil from Corn Ethanol Production via Enzymatic Route," the result of cooperation between a university (COPPE/UFRJ), development agencies (CNPq), and the productive sector (Cesbra), demonstrating the importance of partnerships for the advancement of sustainable solutions	Encourage effective partnerships between the public sector, private sector, and civil society

8 Emissions and Environmental Impacts of Biodiesel in the Life Cycle

After identifying the SDGs covered by the production of biodiesel from residual oil derived from corn ethanol (DCO), it is essential to assess the environmental impacts and emissions associated with the life cycle of this biofuel. Considering the growing concern with air pollution and its effects on public health and the environment, it is increasingly urgent to replace fossil fuels with cleaner and more sustainable energy alternatives [30].

To illustrate the environmental benefits of partially or totally replacing fossil diesel, we highlight the study published by [31]. This study conducted a life cycle assessment (LCA) of biodiesel in the Brazilian market, focusing on quantifying carbon dioxide (CO_2) emissions based on data from the Ministry of Mines and Energy (Ministério de Minas e Energia – MME, in Portuguese) and stoichiometric calculations that considered both direct and net emissions, taking into account carbon sequestration by photosynthesis during oilseed cultivation.

The results showed that biodiesel, when considered throughout its entire life cycle, provides a significant reduction in CO_2 emissions compared to fossil diesel. In 2009, the total replacement of diesel with biodiesel could have represented a reduction of up to 78.5% in net CO_2 emissions. In addition, the progressive increase in the mixture of biodiesel and diesel (such as B5, B20, and B100) proved effective in reducing atmospheric pollutants such as carbon monoxide (CO), particulate matter (PM), and hydrocarbons, although an increase in nitrogen oxide (NOx) emissions was observed with the use of pure biodiesel.

These findings are particularly relevant to the present study, which proposes the use of DCO, a low-cost by-product with no nutritional value, as a raw material for the production of biodiesel via an enzymatic route. As it is an industrial waste product from corn ethanol production, the use of DCO avoids competition with food crops and contributes to the valorization of agro-industrial waste, promoting a circular economy.

In addition to environmental and economic advantages, biodiesel produced from DCO also maintains desirable characteristics such as biodegradability, low toxicity, safe handling, and the generation of glycerin as a high value-added by-product. However, similar to the challenges pointed out by [31], it is important to consider that the scalability of DCO use requires caution regarding the availability of raw materials and large-scale economic viability.

It is therefore clear that the use of DCO for biodiesel production represents a promising alternative that is aligned with environmental sustainability, contributing to climate change mitigation and improved air quality, especially in densely populated urban areas.

9 Concluding Remarks

This study conducted a critical and comparative analysis of the main catalytic routes for biodiesel production—homogeneous (acidic and basic), heterogeneous, and enzymatic—based on a structured review of scientific and technical literature. The analysis encompassed reaction mechanisms, operational parameters, environmental implications, and technological challenges, with the aim of identifying the most promising pathways for sustainable biodiesel synthesis.

Among the evaluated routes, enzymatic catalysis emerged as a particularly attractive alternative, especially due to its ability to operate under mild conditions, reduce energy consumption, minimize the formation of undesirable by-products, and achieve high conversion efficiencies. These advantages are especially relevant when working with raw materials containing high levels of free fatty acids, for which enzymatic processes eliminate the need for pretreatment steps commonly required in conventional methods.

Within this context, Distillers Corn Oil (DCO) was examined as a representative example of a non-edible, low-cost feedstock derived from the corn ethanol industry. Its utilization not only avoids competition with food production but also contributes to waste valorization and circular economy principles. Studies reviewed in this work demonstrate that DCO can be effectively converted into biodiesel via enzymatic catalysis, achieving conversion rates above 90% under optimized conditions.

From a sustainability perspective, biodiesel production via enzymatic routes—especially when based on agro-industrial residues such as DCO—aligns with multiple Sustainable Development Goals (SDGs), including clean energy generation, climate change mitigation, and responsible resource use. Life Cycle Assessment (LCA) results further support the environmental benefits of replacing fossil fuels with biodiesel, highlighting significant reductions in greenhouse gas emissions across the fuel's lifecycle.

In conclusion, while the enzymatic route still faces technical and economic challenges for large-scale implementation, the findings of this study reinforce its potential as a viable, cleaner, and more efficient alternative to traditional chemical catalysis. Continued investment in research and innovation is essential to advance biocatalyst development, optimize operational parameters, and promote the integration of enzymatic biodiesel production into existing industrial infrastructures.

References

1. Hiroshi M et al (2025) Biodiesel in Brazil: reflections on the potential of the main raw materials. Brasilia, DF: EMBRAPA
2. Talamini DJ et al (2022) Description_ feasibility of using corn DDGS and DDG in chicken and pig feeding in Santa Catarina_. EMBRAPA, n. Technical Communication, p 594
3. Geris R et al (2007) Soybean biodiesel-tranesterification reaction for practical classes in organic chemistry new chemistry. [s.l: s.n.]
4. Ramos LP et al (2011) Biodiesel production technologies. Rev Virtual de Quimica 3(5):385–405
5. Pinto AC et al (2005) Biodiesel: an overview. J Braz Chem Soc. [s.l: s.n.]. Disponível em: www.scielo.br
6. Brahma S et al (2022) Biodiesel production from mixed oils: a sustainable approach towards industrial biofuel production. Chem Eng J Adv 10:100284
7. Syakirah Talha N, Sulaiman S (2016) Overview of catalysts in biodiesel production. 11, 1
8. Rezende MJC et al (2021) Biodiesel: an overview II. J Braz Chem Soc Sociedade Brasileira de Quimica
9. Veljković VB et al (2018) Biodiesel production from corn oil: a review. Renew Sustain Energy Rev 91:531–548
10. Ma F, Hanna MA (1999) Biodiesel production: a review. Bioresour Technol 70(1):1–15
11. Vicente Crespo G (2001) Study of the biodiesel production process. [s.l: s.n.].ARGUS MEDIA. Brazil defends ethanol as a maritime fuel. Biofuels, Petrochemicals—Latest market news, October 19, 2023. Available at: https://www.argusmedia.com/pt/news-and-insights/latest-market-news/2500736-brasil-defende-etanol-como-combustivel-maritimo. Accessed on: June 23, 2025
12. Souza VH et al (2014) Study of enzymatic production of biodiesel from residual oil. Eclet Quim 39(1):200–215
13. Zhang Y et al (2003) Biodiesel production from waste cooking oil: 1. Process design and technological assessment. Bioresour Technol 89(1):1–16
14. Leung DYC, Wu X, Leung MKH (2010) A review on biodiesel production using catalyzed transesterification. Appl Energy 87(4):1083–1095
15. Demirbas A (2009) Biodiesel from waste cooking oil via base-catalytic and supercritical methanol transesterification. Energy Convers Manage 50(4):923–927. https://doi.org/10.1016/j.enconman.2008.12.023
16. Marchetti JM, Miguel VU, Errazu AF (2007) Possible methods for biodiesel production. Renew Sustain Energy Rev 11(6):1300–1311
17. Wancura JHC et al (2023) Demystifying the enzymatic biodiesel: How lipases are contributing to its technological advances. Renew Energy 216:119085
18. Fantinel A (2016) Análise Da Produção De Biodiesel Em Pequena Unidade De Processamento Empregando Óleos Residuais Como Matéria-Prima
19. Cao W, Han H, Zhang J (2005) Preparation of biodiesel from soybean oil using supercritical methanol and co-solvent. Fuel 84(4):347–351

20. Szczesna Antczak M et al (2009) Enzymatic biodiesel synthesis—key factors affecting efficiency of the process. Renew Energy 34(5):1185–1194
21. Sharma YC, Singh B, Upadhyay SN (2008) Advancements in development and characterization of biodiesel: a review. Fuel 87(12):2355–2373
22. Gu J et al (2015) Studies on biodiesel production from DDGS-extracted corn oil at the catalysis of Novozym 435/super absorbent polymer. Fuel 146:33–40
23. Mata TM et al (2012) Biodiesel production from corn oil via enzymatic catalysis with ethanol. Energy Fuels Anais 26(5):3034–3041
24. Moreau RA, Hums ME (2020) Corn oil and distillers corn oil. In: Bailey's industrial oil and fat products. [s.l: s.n.], p 1–27
25. Singh N, Cheryan M (1998) Extraction of oil from corn distillers dried grains with solubles. Trans ASAE 41(6):1775–1777
26. Li H et al (2015) Study on the parameters of extrusion pretreatment of corn germ with semi-wet milling in solvent oil extraction. Cereal Chem 92(4):411–417
27. Everton SS et al (2023) The role of Brazil in the advancement of enzymatic biodiesel production. Braz J Chem Eng 40(1):67–80
28. Ciftci ON, Temelli F (2011) Continuous production of fatty acid methyl esters from corn oil in a supercritical carbon dioxide bioreactor. J Supercrit Fluids 58(1):79–87
29. Ciftci ON, Temelli F (2013) Enzymatic conversion of corn oil into biodiesel in a batch supercritical carbon dioxide reactor and kinetic modeling. J Supercrit Fluids 75:172–180
30. Saleh HM, Hassan AI (2024) The challenges of sustainable energy transition: a focus on renewable energy. Appl Chem Eng 7(2):2084
31. Pivotto F, Ferreira ML, Ferreira APNL (2011) Life cycle analysis of biodiesel in the Brazilian market and quantification of emissions released by the use of this fuel. Exacta 9(3):293–300
32. ANP (2025) Biofuels—National Agency of Petroleum, Natural Gas and Biofuels. Available at: https://www.gov.br/anp/pt-br/assuntos/qualidade-de-produtos/biocombustiveis. Accessed on August 1, 2025
33. Zambare V et al (2021) Recent advances in feedstock and lipase research and development towards commercialization of enzymatic biodiesel. Processes MDPI, 1
34. BIOEN—FAPESP Bioenergy Research Program (2025). Biofuels as an immediate and effective solution for the decarbonization of transportation. São Paulo: FAPESP
35. BRAZIL Ministry of Mines and Energy (2022) Report of the subcommittee on marine fuels—fuels of the future. Brasília: MME. Available at: https://www.gov.br/mme/pt-br/programa-combustivel-do-futuro/relatorio_subcomite_combustiveis_maritimos.pdf. Accessed on June 23, 2025
36. EPE—Energy Research Company (2024) Fact Sheet—Decarbonization of Waterway Transportation. Brasília: EPE
37. de Assis TF, Ricci LM, Monteiro TGM, de Abreu VHS, D'Agosto MDA, Santos AS (2022a) Sustainable transport indicators and mitigation actions applied to the green bond principles. In: Carbon footprints of manufacturing and transportation industries. Springer Nature Singapore, Singapore, pp 139–169
38. De Abreu VHS, Da Costa MG, Da Costa VX, De Assis TF, Santos AS, D'Agosto MDA (2022) The role of the circular economy in road transport to mitigate climate change and reduce resource depletion. Sustainability 14(14):8951
39. de Assis TF, Monteiro TGM, de Abreu VHS, D'Agosto MDA, Santos AS (2022b) Enabling the green bonds market for sustainable transport projects based on the measure/monitoring, reporting and verification method. In: Carbon footprints of manufacturing and transportation industries. Springer Nature Singapore, Singapore, pp 1–24
40. de Abreu VHS, da Costa MG, Assis TFD, D'Agosto MDA, Rocha RS, de Paula LOD, Laissone AMS (2025) Assessing the feasibility of enzymatic biodiesel production using the 5W2H framework: a Brazilian case study with distiller's corn oil. Energies 18(20):5460
41. Santos AS, de Abreu VHS, de Assis TF, Ribeiro SK, Ribeiro GM (2020) An overview on costs of shifting to sustainable road transport: a challenge for cities worldwide. Carbon Footpr Case Stud: Munic Solid Waste Manag, Sustain Road Transp Carbon Sequestration 93–121

42. Fukuda H, Kondo A, Noda H (2001) Biodiesel fuel production by transesterification of oils. J Biosci Bioeng 92(5):405–416. https://doi.org/10.1016/S1389-1723(01)80288-7
43. Geris R, Santos NACD, Amaral BA, Maia IDS, Castro VD, Carvalho JRM (2007) Biodiesel de soja: reação de transesterificação para aulas práticas de química orgânica. Química nova 30:1369–1373
44. VICENTE CRESPO G (2004a) Estudio del proceso de producción de biodiesel
45. UNICEF (2025) Millennium development goals. Available at: https://www.unicef.org/brazil/objetivos-de-desenvolvimento-sustentavel. Accessed Aug 1 2025

Role of Ethanol as an Alternative for the Sustainable Expansion of Electricity Supply in Brazil

Daniel Michilini Carocha and Victor Hugo Souza de Abreu

Abstract This chapter investigates the role of ethanol as an alternative for the sustainable expansion of electricity supply in Brazil. The research contextualizes the growing demand for renewable energy sources and the challenges of greenhouse gas emissions. The objectives include analysing the viability of ethanol as a replacement for fossil fuels and assessing its environmental and economic impact. The methodology involves a literature review and analysis of production and consumption data. The main results indicate that ethanol is an alternative and still contributes to the reduction of emissions from the energy matrix, and its large-scale implementation will not face technical and economic barriers, with new generation technologies that have already overcome the challenges. The study concludes with recommendations for public policies that encourage the use of ethanol as a fuel in the electricity sector.

Keywords Ethanol · Decarbonization · Electric sector · Renewable energy · Public policies

1 Introduction

Brazil is internationally recognized for having one of the cleanest energy matrices, with a substantial share of renewable sources, predominantly hydroelectric generation. However, the increasing penetration of non-controllable renewable sources,

D. M. Carocha
MBA in Energies from the Alberto Luiz Coimbra Institute for Graduate Studies and Engineering Research at the Federal University of Rio de Janeiro, Rio de Janeiro, Brazil
e-mail: danielcarocha@yahoo.com.br

V. H. S. de Abreu (✉)
Civil Engineering at the Polytechnic School of the Federal University of Rio de Janeiro, Postdoctoral Fellow in Transportation Engineering at the Alberto Luiz Coimbra Institute for Graduate Studies and Engineering Research at the Federal University of Rio de Janeiro, Rio de Janeiro, Brazil
e-mail: victor@pet.coppe.ufrj.br; vhsa@poli.ufrj.br

S. S. Muthu (ed.), *Energy Efficiency and Carbon Footprint Assessment*, Environmental Footprints and Eco-design of Products and Processes,
https://doi.org/10.1007/978-3-032-10909-5_7

such as solar and wind power, into the National Interconnected System (Sistema Interligado Nacional—SIN, in Portuguese) has introduced new operational challenges due to their intermittency. The resulting changes in the net load profile, increasingly resembling the "duck curve", require flexible and dispatchable resources capable of meeting peak demand and compensating for sudden fluctuations in generation, thus increasing the complexity of system operation for the National System Operator (Operador Nacional do Sistema Elétrico—ONS, in Portuguese).

To ensure the continuity and security of energy supply in the country, the recent Capacity Reserve and Reserve Energy auctions contracted a considerable amount of supply solutions based on fossil sources. As shown in Graph 1, LRCAP 2021 (Capacity Reserve Auction), PCS 2021 (Simplified Competitive Procedure) and LRCE 2022 (Capacity Reserve Auction in the form of Energy) together contracted a total of 7.8 GW of power from projects that use natural gas and fuel oil as energy sources, according to [1].

These measures could hinder the country's energy transition policy, which is advocated both in the Paris Agreement (COP 21) and in the National Plan on Climate Change (Plano Clima), according to [2, 3]. These initiatives have as their main objective to guide Brazil's actions in reducing greenhouse gas emissions and combating the impacts of climate change, with broad advocacy of increasing renewable energy capacity and the development of biofuels.

From this perspective, ethanol emerges as a renewable biofuel with a long-established production chain in Brazil, capable of offering dispatchable, flexible, and low-carbon generation for the electricity sector. While its role in the transportation sector is well documented, its potential for integration into thermoelectric generation—especially to complement intermittent sources and ensure

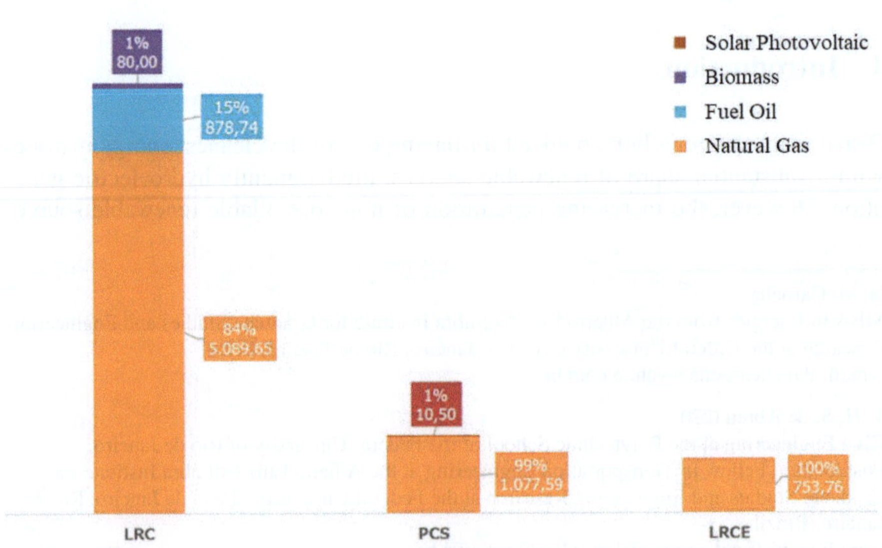

Graph 1 Energy contracted by source in MW [1]. *Source*

peak supply—remains underexplored in both the literature and national policy frameworks.

In this context, this study aims to highlight ethanol in Brazil as a potential sustainable solution in offering reliability to the electricity sector, especially for peak service. We present how this alternative is effective in guaranteeing energy supply to SIN, with the flexibility necessary for system operation, and how its adoption can align with the country's decarbonization targets. The analysis combines a targeted literature review with quantitative assessment of production, consumption, emissions, and cost data, providing an integrated technical, environmental, and regulatory perspective.

2 Methodology

This study was developed using a qualitative and quantitative approach, structured in two main stages: (i) systematic and targeted literature review; and (ii) exploratory analysis of secondary data on ethanol production, consumption, and potential applications in the Brazilian electricity sector.

2.1 Literature Review

The literature review focused on identifying technical-scientific evidence and institutional data to support the assessment of ethanol as a sustainable alternative for the expansion of electricity supply in Brazil. Indexed academic publications, sectoral technical reports, legal regulations, and strategic policy documents were included, prioritizing materials published between 2008 and 2024.

Key sources included reports from the Energy Research Company (Empresa de Pesquisa Energética—EPE, in Portuguese), the Electric Energy Commercialization Chamber (Câmara de Comercialização de Energia Elétrica—CCEE, in Portuguese), the Ministry of Mines and Energy (Ministério de Minas e Energia—MME, in Portuguese), the National Agency for Petroleum, Natural Gas and Biofuels (Agência Nacional do Petróleo, Gás Natural e Biocombustíveis—ANP, in Portuguese), and relevant industry organizations such as Raízen and Wärtsilä. The research considered keywords such as *ethanol fuel*, *biofuel-based thermoelectric generation*, *operational flexibility*, *capacity reserve*, and *electricity matrix decarbonization*.

The screening process prioritized documents providing quantitative data, comparative analyses with fossil fuels, and descriptions of technologies applicable to ethanol-based electricity generation, especially stationary Otto-cycle internal combustion engines.

2.2 Secondary Data Analysis

The data analysis consisted of collecting, consolidating, and interpreting information from public databases and sectoral reports. The following were used:

- Ethanol production and consumption data [4, 10];
- Statistics on the share of energy sources in the transportation sector and in the domestic energy supply [5];
- Technical characteristics and performance of ethanol-based power generation technologies, obtained from manufacturer specifications (e.g., Wärtsilä 20V32 platform);
- Greenhouse gas (GHG) emission factors for ethanol, diesel oil, fuel oil, and natural gas, obtained from standardized sources such as ACV Raízen 23′24, GHG Protocol, Renovacalc, and RED II.

To estimate potential emission reductions, emission factors expressed in kg CO_2/MWh were combined with average thermoelectric plant efficiencies, as provided by the *Ecoinvent 3.10* database ("electricity production, oil—BR-Southern grid—electricity, high voltage"). Projections considered an annual operating regime of 7,800 h and an average power output of 6.5 GW per thermoelectric unit.

The data were comparatively analyzed to contrast the environmental and operational performance of ethanol with that of the main fossil fuels used in the national thermoelectric park (natural gas, diesel oil, and fuel oil). This comparison made it possible to estimate the magnitude of CO_2 e reductions achievable through the partial or total replacement of these fuels with ethanol.

2.3 Integration of Analyses

The evidence gathered from the literature review and quantitative analyses was integrated to assess:

- The technical feasibility of ethanol-based thermoelectric generation within the context of the SIN;
- The potential contribution to Brazil's decarbonization targets under the Paris Agreement and its Nationally Determined Contribution (NDC);
- Compatibility with operational requirements for flexibility and fast dispatch capacity defined in Capacity Reserve Auctions (LRCAP).

This methodology provided a robust technical basis for discussing the strategic role of ethanol in the electricity sector, considering not only its availability and technological maturity but also its comparative advantages over fossil-based solutions in terms of energy security, operational flexibility, and environmental performance.

3 Results

The production of energy from sugarcane biomass plays a strategic role in the national energy supply, establishing the sugar and alcohol sector as one of the key contributors to Brazil's energy matrix. Ethanol, in addition to being a renewable alternative to fossil fuels, also stands out for its potential in electricity generation, contributing to supply diversification and the reduction of greenhouse gas emissions. In this context, this section examines the participation of the sugar and alcohol sector in the internal energy supply, as well as the technologies used in ethanol-based electricity generation and their contribution to the decarbonization of the energy system.

3.1 Current Role of Ethanol in Brazil and Prospects for the Electricity Sector

In this scenario, ethanol is a sustainable national biofuel widely used to supply energy in the transportation sector. Since the implementation of the National Alcohol Program (Proalcool) in the 1970s, the ethanol industry has established itself in Brazil as a sustainable and competitive alternative, contributing to the security of supply and the decarbonization of the country's energy matrix.

In Brazil, ethanol is produced mainly from sugar cane and corn, according to [4], and is the third most used fuel in the country in the transport sector, accounting for 17% of total energy consumption in this segment, behind only diesel and gasoline, according to [5], the according to Graph 2.

Ethanol can play a strategic role in the Brazilian electricity sector, through an innovative dispatchable, flexible and low-carbon thermoelectric generation solution. This solution has technical characteristics that meet the capacity, reliability and flexibility requirements defined for the 2025 Capacity Reserve Auction in the form of Power (LRCAP). However, for biofuel-based solutions to become a reality, it is necessary to improve the auction rules and the clauses of the capacity reserve contracts, to provide competitive conditions for this source.

3.2 Participation of the Sugar and Alcohol Sector in the Internal Energy Supply

According to the 2024 National Energy Balance (Chart 3), the sugar and alcohol sector was responsible for 17% of the internal energy supply in Brazil, becoming the second largest energy source in the country, behind only oil and its derivatives, according to [5].

The sugar and alcohol sector contributes to energy supply through ethanol, used mainly in the light vehicle transportation sector, and through the cogeneration of

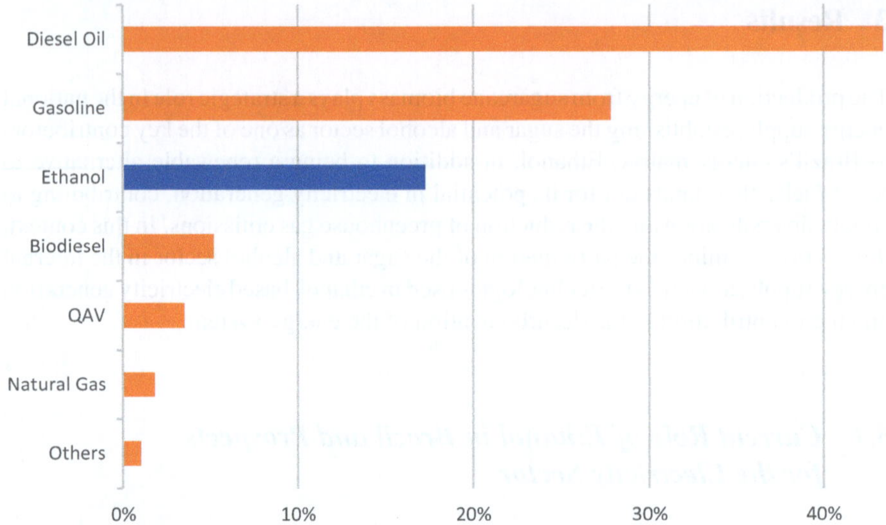

Graph 2 Participation of energy sources in transport [5]. *Source*

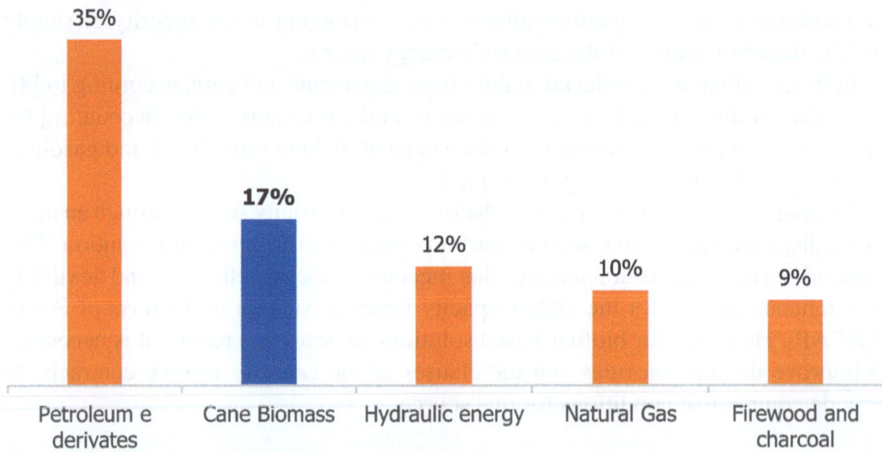

Chart 3 Internal energy supply 2022–2023 [5]. *Source*

electricity and heat from sugarcane bagasse, which is used in the production of ethanol and sugar. The heat and part of the electricity generated are consumed internally in the production process, while the surplus electricity is exported to the National Interconnected System.

In addition to these already established processes, the sugar and alcohol industry is expanding its energy supply by increasing the efficiency of cogeneration plants and by burning sugarcane tops and straw together with bagasse to produce biomethane and second-generation ethanol (E2G).

A study conducted by the institutions Agroicone, Universidade Estadual de Campinas (Unicamp) and the Brazilian Agricultural Research Corporation (Embrapa) highlights the positive impact of sugarcane production on climate change mitigation in Brazil over the last two decades. Land use dynamics and the evolution of sugarcane cultivation practices have resulted in the removal of approximately 9.8 million tons of CO_2 from the atmosphere annually, according to [6].

3.3 *Ethanol-Based Electricity Generation Technology and Its Decarbonization Potential*

A significant advance for the Brazilian electricity sector is the use of ethanol-powered engines, based on an established platform that allows generating up to 10 MW of power per unit, such as the 20V32 platform already available on the market.

Ethanol engines can be effectively applied in Brazil, as they are designed to offer reliability, efficiency and versatility in generation, aiding in the integration of renewable energies. In periods of water scarcity, such as those that occurred in 2020–2021, these engines can operate continuously, significantly reducing net greenhouse gas emissions compared to fossil sources.

Furthermore, under normal hydrological conditions, these engines can operate intermittently and in a controllable manner, complementing short-term variations in wind and solar generation, in addition to meeting peak demands. Standing out for its flexibility, this ethanol-powered technology can reach full power in less than 10 min from a cold start and shut down from full power in about 1 min.

The advantages of the ethanol solution are not limited to the technological aspects of the engines, but also include the fuel, whose supply logistics are already well established and presents advantages over natural gas, especially in the context of the dispatchable generation flexibility currently required by the Brazilian electricity sector. In addition to the technical aspects of the fuel, its potential strategic role in the decarbonization of the Brazilian electricity sector and in the country's energy sovereignty stands out.

Given Brazil's commitment to reduce its greenhouse gas emissions by 53% by 2030 and achieve carbon neutrality by 2050, according to MMA: Nationally Determined Contribution (NDC) [7], the country faces the need to expand its thermal park to ensure energy security. Brazil currently has an installed capacity of around 20 gigawatts (GW) in thermal power plants, and, according to the government plan, this capacity should be doubled by 2031, according to according to [8]. However, the expansion of this park cannot happen at the expense of fossil fuels. In this context, ethanol emerges as a strategic and sustainable alternative.

The use of ethanol as a fuel in thermal power plants, compared to diesel, natural gas or fuel oil, results in a significant reduction in greenhouse gas emissions. To illustrate, a thermal power plant powered by ethanol emits approximately 79% less CO_2 than one powered by diesel or heavy fuel oil, and 74% less than one powered by natural gas. This positive impact can be quantified: by replacing a 5–8 GW thermal power plant powered by natural gas with one powered by ethanol, the emission of up to 3.8 million tons of CO_2 per year is avoided. If the replacement is of a diesel or heavy fuel oil plant, this savings amounts to approximately 5 million tons of CO_2 per year.

The calculation of CO_2 emissions is based on the following sources:

- Ethanol Emission Factor: ACV Raízen 23′24—Hydrated Ethanol 1G; GHG Protocol v2024.0.2—Hydrated Ethanol—emission factors for stationary combustion.
- Diesel Oil Emission Factor: Renovacalc v.7; GHG Protocol v2024.0.2—Pure Diesel Oil—emission factors for stationary combustion.
- Natural Gas Emission Factor: Renovacalc v.7; GHG Protocol v2024.0.2—Wet Natural Gas—emission factors for stationary combustion.
- Fuel Oil Emission Factor: RED II; GHG Protocol v2024.0.2—Fuel Oil—emission factors for stationary combustion.

The average efficiency of thermoelectric plants (kg/KWh) was estimated based on the Ecoinvent 3.10 dataset ('electricity production, oil—BR-Southern grid—electricity, high voltage'). The annual operating period considered was 7,800 h, with an average power of 6.5 GW per thermoelectric unit (Graph 4).

Additionally, in economic terms, according to data and information from the Ministry of Mines and Energy, ANP and Petrobras [9–11], although the generation cost per megawatt-hour of ethanol is higher compared to fossil fuels, its decarbonization advantage outweighs this difference (Chart 5).

Considering the significant reductions provided by ethanol fuel, which are about four times greater than those of fossil alternatives, the cost differential becomes less relevant compared to the enormous decarbonization benefits. In other words, although

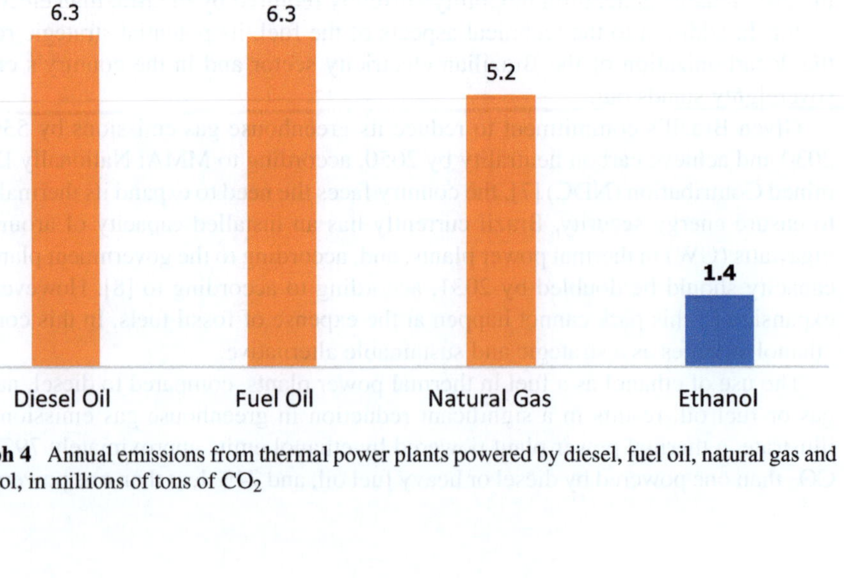

Graph 4 Annual emissions from thermal power plants powered by diesel, fuel oil, natural gas and ethanol, in millions of tons of CO_2

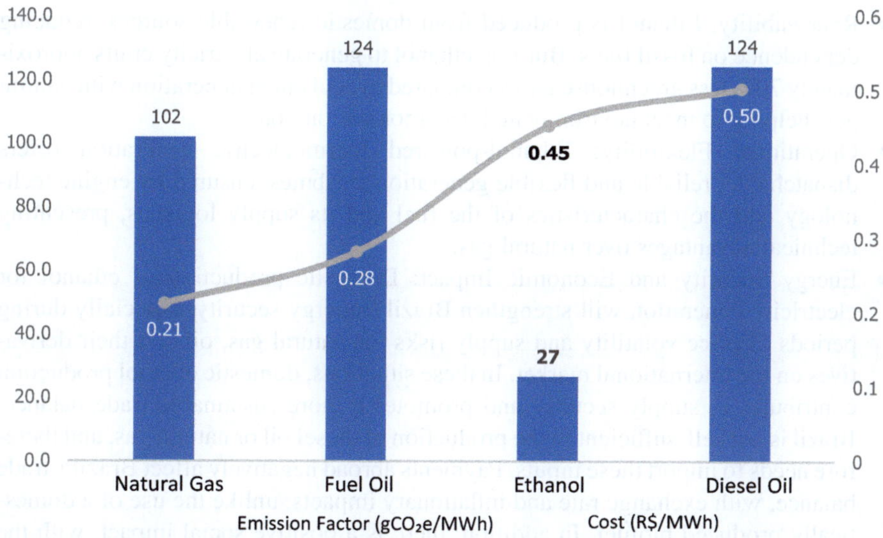

Chart 5 Cost per MWh versus fuel emission factor

the cost of ethanol is approximately twice that of natural gas, it offers emissions reductions up to four times greater, making it an extremely effective solution for decarbonizing the energy sector.

Brazil currently has an installed capacity of around 20 gigawatts (GW) in thermoelectric plants, and, according to the government plan, this capacity should be doubled by 2031, according to [8]. If the country doubles the capacity of its thermal park using exclusively ethanol, approximately 15 million tons of CO_2 per year can be avoided, representing 6.2% of the total reduction in emissions, for the energy sector, that Brazil needs to achieve by 2030 to meet its commitments established in the NDC (Nationally Determined Contribution).

Given the current context, the expansion of thermoelectric plants in Brazil must focus on solutions that balance energy security with the reduction of greenhouse gas emissions. In other words, ethanol is the biofuel capable of replacing fossil fuels, such as natural gas and diesel oil, to achieve the objectives of transitioning the energy matrix towards a sustainable environment, which is a right of citizens and a responsibility of the State.

4 Discussion

Initial results suggest that the use of ethanol as a fuel can result in a significant reduction in greenhouse gas emissions. In addition, it presents favorable economic potential for both industry and farmers, strengthening Brazil's energy sovereignty.

We highlight the strong deliverables:

- Renewability: Ethanol is produced from domestic renewable sources, reducing dependence on fossil fuels. Burning ethanol to generate electricity emits approximately 70% less greenhouse gases compared to equivalent generation with natural gas, helping to meet environmental decarbonization goals.
- Operational Flexibility: Ethanol-powered thermoelectric generation offers dispatchable, reliable and flexible generation attributes, ensured by engine technology and the characteristics of the fuel and its supply logistics, presenting technical advantages over natural gas.
- Energy Security and Economic Impact: Domestic production of ethanol for electricity generation will strengthen Brazil's energy security, especially during periods of price volatility and supply risks for natural gas, oil and their derivatives on the international market. In these situations, domestic ethanol production contributes to supply security and promotes a more sustainable trade balance. Brazil is not self-sufficient in the production of diesel oil or natural gas, and therefore needs to import these inputs. Payments abroad negatively affect Brazil's trade balance, with exchange rate and inflationary impacts, unlike the use of a domestically produced biofuel. In addition, there is a positive social impact, with the generation of jobs in Brazil throughout the ethanol production chain, in contrast to the limited generation of jobs in the country for imported inputs.

The defense and preservation of an ecologically balanced environment for present and future generations is a precept established in section VIII of § 1 of article 225 of the Federal Constitution (CF/88), included by Constitutional Amendment (EC) 123/ 2022. The provision states:

"Art. 225. Everyone has the right to an ecologically balanced environment, a common good for the people and essential to a healthy quality of life, imposing on the Public Authorities and the community the duty to defend and preserve it for present and future generations.

§ 1° To ensure the effectiveness of this right, it is the responsibility of the Public Authorities:

VIII—maintain a favorable tax regime for biofuels and low-carbon hydrogen, in the form of a supplementary law, in order to ensure that they are subject to lower taxation than that levied on fossil fuels, capable of guaranteeing a competitive advantage in relation to the latter, especially in relation to the contributions referred to in art. 195, I, "b", IV and V, and art. 239 and the taxes referred to in arts. 155, II, and 156-A."

It is clear that the constitutional determination, which is immediately applicable, applies to all Public Authorities. Likewise, a differentiated tax regime must include all tax instruments available to the Public Authorities, such as taxes, credits, incentives, subsidies, financing, or any other way of ensuring the competitiveness of biofuel in relation to the fossil fuel it will replace. This provision remains valid, including in the new tax system introduced by Constitutional Amendment No. 132/2023.

In this context, the constituent's intention to seek, guarantee and promote the sustainable development of the country is clear. Among the various measures that the Government is responsible for achieving this objective, the need to adopt a sustainable public procurement policy stands out. We understand that power capacity auctions, currently responsible for meeting supply security criteria, should include solutions based on ethanol as a fuel. However, for these solutions to be able to compete in auctions and contribute efficiently to the Brazilian electricity system, regulatory improvements are needed in the auction rules and contracts, allowing biofuel to be competitive.

4.1 Challenges in MME Ordinances

First, it is important to emphasize that the proposed solution already allows ethanol-fired thermoelectric plants to meet the operational flexibility requirements established in MME Ordinance No. 774/2024. However, it is essential that the competitiveness conditions of this renewable and national fuel include adjustments indexed to market prices, monthly, similar to what occurs with fossil fuels.

Currently, the ethanol solution is classified by ANEEL as biomass, with fuel costs indexed to the IPCA and annual adjustments, according to MME Ordinance No. 42/2007. However, ethanol follows its own commodity price dynamics, distinct from the IPCA, influenced by variables such as supply and demand, production and logistics costs, and tax burden.

Thus, the risk of volatility in generation costs associated with fuel prices would be practically incalculable for a potential ethanol generator, especially in long-term contracts (PPA) with no provision for renegotiating economic clauses.

This is not the case with fossil-fueled generation solutions, such as natural gas, which can index their costs to a basket of international prices, the US dollar and the IPCA, depending on the generator's choice. In this case, generators can mitigate the risks associated with fuel price volatility by indexing their electricity contracts to their fuel supply contracts, establishing a financial "hedge" between revenues and expenses.

Therefore, it is crucial to review MME Ordinance No. 42/2007, which defines the conditions for indexing and readjusting generation costs associated with fuel prices in energy contracts and capacity reservations. Reviewing this regulation is essential to ensure the economic viability of ethanol solutions for the National Interconnected System.

These regulatory updates are essential to ensure balanced competitive conditions for ethanol, which will be a sustainable solution for generating electricity in the Brazilian market, aligned with the country's environmental energy transition policies towards a low-carbon economy.

4.2 Limits for Unit Variable Cost (UVC)

Regarding the Unit Variable Cost (UVC), it is crucial that the UVC ceiling can accommodate the different types of fuels in the 2027 and 2028 Thermoelectric Power products. This can be done through a single UVC ceiling that is sufficient for all fuels or by adopting a specific UVC ceiling for each fuel type.

To increase competitiveness in the auction, with diversified supply solutions and business models, it is necessary that the CVU ceiling to be established considers the price levels of the different fuels (molecules), including biofuels, as well as the different logistics cost structures associated with the fuel supply.

4.3 Creation of the Product of Renewable Thermal Power Plants

To ensure the competitiveness of biofuels in relation to fossil fuels, it is essential to create a specific power product for renewable thermoelectric plants, to be included in the products negotiated in the 2025 LRCAP.

Brazil has a significant history of providing incentives to promote renewable energy, aiming to diversify its energy matrix and reduce dependence on fossil fuels. An important example is the Program for the Incentive of Alternative Sources of Electric Energy (Proinfa), created in 2002, one of the first major Brazilian programs aimed at stimulating alternative sources of electricity.

Some other programs and policies have fostered the development of the sector over the years. For example, the energy auction system, which began to include specific categories for renewable sources in 2007. These actions demonstrate how government action, through subsidies and incentives, can generate positive results for the country.

Furthermore, the potential for job growth that the incentive for renewable thermal power plants can generate must be considered. According to the government plan mentioned above, the installed capacity of thermal power plants should double by 2031. Based on ethanol-powered thermal power plants, it is estimated that 11,300 jobs will be created within the parks alone, not counting the entire production chain, which can increase this number even further.

It is also important to highlight the potential for replicating this technology in other countries, strengthening the foreign ethanol market. The operational complexity of meeting peak demand, as seen in Brazil, also affects other countries undergoing energy transition with the increase in intermittent renewable sources in their matrices.

Given all these points, we understand that 5% of the total power to be contracted in the 2025/26 LRCAP must be specifically allocated to renewable thermoelectric plants.

4.4 Recommendations

In view of the above, we bring three recommendations so that renewable and flexible thermal plants powered by ethanol become a competitive option in power contracting competitions.

The first suggestion is that a change be made to Art. 3, §1, of MME Ordinance No. 42/2007, with the inclusion of item VII, as described below:

"VII-for thermoelectric projects powered by biofuels, it P_v will be, for each month "M", given according to the following expression:

$$P_v = a \times P_{B100i} + b \times P_{e\tan olNE} + d$$

- a, b, c: parameters established by the entrepreneur when registering the enterprise for Technical Qualification with the EPE;
- $d =$ parameter established by the entrepreneur when registering the enterprise for Technical Qualification with the EPE, updated annually by the INPC of fuel ethanol;
- $P_{B100_i} =$ Weekly price of biodiesel (B-100) in Brazil, published by the National Agency of Petroleum, Natural Gas and Biofuels (ANP), simple average of the last 4 weeks for the i reference region of origin of the renewable. Values in R$/liter, net of taxes and freight;
- $P_{etanol_{i,CSNO}} =$ Price of fuel ethanol i (anhydrous or hydrated) in the State of São Paulo, published by CEPEA/ESALQ, in month "M-1", for renewable originating in the Center-South (CS) and North (NO) regions of Brazil. Values in R$/liter, net of taxes and freight, in the PVU condition (station vehicle plant);
- $P_{etanol_{NE}} =$ Price of anhydrous fuel ethanol Suape (Anhydrous NNE Brazil delivered Suape), as published in the Platts Fuel Ethanol Report, for renewable fuel originated in the Northeast (NE) region of Brazil. Simple average of the last 4 weeks. Values in R$/liter, net of taxes and freight, in DAP (delivered at place) condition.

The second suggestion is that a CVU ceiling be established that takes into account the price levels of different fuels, including biofuels, as well as the different logistics cost structures associated with fuel supply. The third suggestion is the creation of a specific power product for renewable thermoelectric plants and that 5% of the total power to be contracted in the 2025 LRCAP be allocated exclusively for this product.

5 Final Considerations

The present study has demonstrated that ethanol plays a strategic role in the sustainable expansion of electricity supply in Brazil. Its use in thermoelectric generation offers significant advantages in terms of operational flexibility, energy security, and the reduction of greenhouse gas emissions. The analysis of production and consumption data indicated that the ethanol supply infrastructure is well established, allowing its integration into the electricity sector without significant technical barriers.

The results indicate that, in addition to contributing to the diversification of the electricity matrix, ethanol can also play a crucial role in reducing dependence on fossil fuels, aligning with the climate goals established by Brazil. The use of ethanol-powered engines in thermoelectric plants proves to be a viable alternative to meet peak demand and ensure stability in the National Interconnected System (SIN), especially given the intermittency of solar and wind sources.

Beyond its environmental advantages, ethanol can strengthen Brazil's energy sovereignty by reducing dependence on imported fossil fuels, mitigating exposure to price volatility in international markets, and fostering domestic job creation along the production chain. These benefits are consistent with the country's Nationally Determined Contribution (NDC) under the Paris Agreement and reinforce its leadership in renewable energy development.

However, for ethanol to be fully integrated into the electricity sector, it is essential to advance public policies that encourage its use. This includes regulatory adjustments in capacity reserve auctions, the creation of financial incentives, and the promotion of research and development of ethanol-based generation technologies. Furthermore, the expansion of second-generation ethanol (E2G) and biomethane production can further strengthen its competitiveness and sustainability.

Thus, ethanol presents itself as a strategic solution for the future of the Brazilian electricity sector, enabling the transition to a cleaner and more resilient energy matrix. The implementation of appropriate measures will allow Brazil to fully harness the potential of this biofuel, consolidating its leadership in renewable energy production and significantly contributing to combating climate change.

Future studies should focus on evaluating the economic viability of large-scale ethanol-based thermoelectric plants, considering different regulatory and market scenarios. Additionally, further research is needed to optimize the efficiency of ethanol combustion in power generation, exploring technological innovations that enhance its performance. Another relevant aspect is the assessment of the environmental and socio-economic impacts of expanding ethanol use in electricity generation, ensuring that its adoption aligns with sustainability principles and energy security objectives.

Acknowledgements I would like to thank the professors and colleagues from Wärstilä (Jorge Alcaide, Kent Westergård, Paulo Castro, Gabriel Cavados, Ville Kallioniemi, Jorge Bezerra, Joice Peixoto) and from Raízen Power (Bianca Lima) who contributed to the completion of this work.

References

1. CCEE (2024) Consolidated auction results report—Reference 08/2024. Available at: https://www.ccee.org.br/acervo-ccee?especie=38753&assunto=39056&keyword=consolidado&periodo=1825
2. MME (2015) Paris agreement. Available at: https://antigo.mma.gov.br/clima/convencao-das-nacoes-unidas/acordo-de-paris.html
3. MME (2008) National plan on climate change. Available at: https://antigo.mma.gov.br/clima/politica-nacional-sobre-mudanca-do-clima/plano-nacional-sobre-mudanca-do-clima.html
4. EPE (2023) Analysis of the biofuels situation—year 2022. Available at: https://www.epe.gov.br/sites-pt/publicacoes-dados-abertos/publicacoes/PublicacoesArquivos/publicacao-756/NT-EPE-DPG-SDB-2023-01_Analise_de_Conjuntura_dos_Biocombustiveis_Ano2022.pdf
5. EPE (2024) National energy balance—base year 2023. Available at: https://www.epe.gov.br/sites-pt/publicacoes-dados-abertos/publicacoes/PublicacoesArquivos/publicacao-819/topico-715/BEN_S%C3%ADntese_2024_PT.pdf
6. RAÍZEN (2024) Integrated report harvest 23'24. Available at: https://raizen.com.br/items-files/item-1212-relatorio-raizen-2023-24-1508.pdf
7. MMA (n.d.) Nationally determined contribution (NDC). Available at: https://www.gov.br/mma/pt-br/assuntos/mudanca-do-clima/NDC
8. Agência Brasil (2024) Brazil is not at risk of an energy crisis in 2024, says minister. Agência Brasil, September 22, 2024. Available at: https://agenciabrasil.ebc.com.br/geral/noticia/2024-09/brasil-nao-corre-risco-de-crise-energetica-em-2024-diz-ministro
9. MME (2024) Natural gas industry monitoring bulletin—April 2024. Available at: https://www.gov.br/mme/pt-br/assuntos/secretarias/petroleo-gas-natural-e-biocombustiveis/publicacoes-1/boletim-mensal-de-acompanhamento-da-industria-de-gas-natural/2024/04-boletim-de-acompanhamento-da-industria-de-gas-natural-abril-de-2024.pdf/view
10. ANP (n.d.) Fuel distribution prices. Available at: https://www.gov.br/anp/pt-br/assuntos/precos-e-defesa-da-concorrencia/precos/precos-de-distribuicao-de-combustiveis
11. Petrobras (2024) Fuel prices. Available at: https://precos.petrobras.com.br/documents/d/precos-dos-combustiveis/tabelas-de-precos-oc-18-09-24-pdf

MIX
Papier aus verantwortungsvollen Quellen
Paper from responsible sources
FSC® C105338

If you have any concerns about our products,
you can contact us on
ProductSafety@springernature.com

In case Publisher is established outside the EU,
the EU authorized representative is:
Springer Nature Customer Service Center GmbH
Europaplatz 3, 69115 Heidelberg, Germany

Printed by Libri Plureos GmbH
in Hamburg, Germany

